CRE▲TIVE
HOMEOWNER®

50 PROJECTS FOR BUILDING YOUR
BACKYARD HOMESTEAD

UPDATED & EXPANDED EDITION

A HANDS-ON, STEP-BY-STEP SUSTAINABLE-LIVING GUIDE

DAVID TOHT

CREATIVE HOMEOWNER®

50 Projects for Building Your Backyard Homestead, Updated & Expanded Edition

Managing Editor: Gretchen Bacon
Acquisitions Editor: Lauren Younker
Editor: Joseph Borden
Designer: Mike Deppen
Proofreader: Kelly Umenhofer
Indexer: Jay Kreider

Printed in China

ISBN 978-1-58011-610-7

Cataloging-in-Publication data is on file with the Library of Congress

We are always looking for talented authors. To submit an idea,please send a brief inquiry to acquisitions@foxchapelpublishing.com.

CREATIVE HOMEOWNER®
www.creativehomeowner.com
Creative Homeowner books are distributed exclusively by
Fox Chapel Publishing
903 Square Street
Mount Joy, PA 17552
www.FoxChapelPublishing.com

DAVID TOHT has more than 60 how-to books to his credit. He considers harvesting a sun-warmed tomato from his own garden one of life's sweetest pleasures. He and his wife, Rebecca, live in Minnesota.

Acknowledgments

Special thanks to those who generously shared their time and expertise, including Luis Apolinar, David and Heather Armitage, Dan Aulwurm, Chuck Bauer, Mike Bergey Marc Bernard, Sylvia Bernstein, Dr. Sally Brown, Kurt Hollis, Daniel Hull, Rebecca Brody Kamerman, Adam Matthews, Roy Stark McGarrah, Sean Millhorn, Ray Rasmussen, John Redd, Wes Shank, Terry and Jennifer Shaw, Ernie Schmidt, Adam Toht, and Robin Forsythe and Becky and Paul Tuttle.

Safety First

Though all concepts and methods in this book have been reviewed for safety, it is not possible to overstate the importance of using the safest working methods possible. What follows are reminders—do's and don'ts for yard work and landscaping. They are not substitutes for your own common sense.

- *Always* use caution, care, and good judgment when following the procedures described in this book.

- *Always* determine locations of underground utility lines before you dig, and then avoid them by a safe distance. Buried lines may be for gas, electricity, communications, or water. Start research by contacting your local building officials. Also contact local utility companies; they will often send a representative free of charge to help you map their lines. In addition, there are private utility locator firms that may be listed in your Yellow Pages or online. Note: previous owners may have installed underground drainage, sprinkler, and lighting lines without mapping them.

- *Always* read and heed the manufacturer's instructions for using a tool, especially the warnings.

- *Always* ensure that the electrical setup is safe; be sure that no circuit is overloaded and that all power tools and electrical outlets are properly grounded and protected by a ground-fault circuit interrupter (GFCI). Do not use power tools in wet locations.

- *Always* wear eye protection when using chemicals, sawing wood, pruning trees and shrubs, using power tools, and striking metal onto metal or concrete.

- *Always* read labels on chemicals, solvents, and other products; provide ventilation; heed warnings.

- *Always* wear heavy rubber gloves rated for chemicals, not mere household rubber gloves, when handling toxins.

- *Always* wear appropriate gloves in situations in which your hands could be injured by rough surfaces, sharp edges, thorns, or poisonous plants.

- *Always* wear a disposable face mask or a special filtering respirator when creating sawdust or working with toxic gardening substances.

- *Always* keep your hands and other body parts away from the business ends of blades, cutters, and bits.

- *Always* obtain approval from local building officials before undertaking construction of permanent structures.

- *Never* work with power tools when you are tired or under the influence of alcohol or drugs.

- *Never* carry sharp or pointed tools, such as knives or saws, in your pockets. If you carry such tools, use special-purpose tool scabbards.

Contents

6 Introduction

10 Garden Structures

12 Building a Wooden Raised Bed
19 Constructing a Concrete Block Raised Bed
27 Irrigating a Raised Bed
30 Building Keyhole Garden
36 Making a Bottom-Watered Container Garden
40 Making a Vertical Planter
44 Accessible Gardens and Paths
46 Building an Inclined Planter
49 Building an Arbor
60 Adding a Trellised Arbor
67 Building a Cucumber Trellis
68 Installing a Tool-Storage Rack
71 Making a Grow Light Stand
75 Making Soil Blocks

78 Fences and Pens

80 Post Foundations
81 Installing Posts
84 Notching Posts
86 Installing a Picket Fence
88 Choosing a Horizontal-Board Fence
89 Installing a Horizontal-Board Fence
90 Installing a Vertical-Board Fence
91 Installing a Wood-and-Wire Fence
93 Choosing Gate Latches
94 Making a Picket Gate
96 Stretching a Fence
100 Adding a Gate
106 Installing a Solar-Powered Electric Fence
111 Making a PVC Hen Pen
119 Making a PVC Hurdle

124 Housing Chickens

124 Building a Chicken Coop and Run
142 Building an A-frame Chicken Tractor
152 Coping with the Cold
156 Prepping for Extreme Heat

158 Building Sheds

160 Building Basics
168 Saltbox Shed
194 Goat Shed
206 Roofing Alternatives
210 Setting Up a Backyard-Homestead Shop

212 Wind and Solar Power

212 Installing a Pump or Aeration Windmill
218 Installing Solar Power
222 Wind Turbines for Electricity
224 How Wind Systems Work
225 Hybrid Systems

226 Aquaponics & Hydroponics

228 Understanding Aquaponics
232 Understanding Hydroponics
233 Assembling a Hydroponic System

236 Building Beehives

236 Building a Langstroth Hive
244 Making a Warré Hive
245 Building a Top Bar Hive

226 Plumbing & Wiring

250 No-Sweat Sweating
251 Water Where You Need It
252 Installing an Anti-Freeze Spigot
254 Repairing a Freeze-Proof Spigot
255 Adding a Freeze-Proof Yard Hydrant
257 Your Friend, the GFCI Receptacle
259 Installing a GFCI Receptacle
260 Running Outdoor Conduit and Cable
261 Outdoor Electrical Boxes
262 Installing Outdoor Receptacles and Switches
263 Conduit and Fittings
264 Installing UF Cable
266 Installing a Stand-Alone Receptacle
267 Adding Supplemental Light for Poultry

269 Resources
268 Index

Introduction

When I was young, my grandparents owned a 240-acre diversified farm in west-central Illinois. For us kids, too young to pitch in with the chores, it was a wonderful playground. If we weren't hanging on the fence staring down steers (with one prodigy always coming forward to have his forehead scratched), we were gingerly reaching under hens for eggs or slapping the dusty backs of piglets. The haymow, redolent of alfalfa, was a wonderful jungle gym for climbing, building forts, or swinging Tarzan-style on the dusty old ropes. Our sandbox was a pile of sawdust—hen-house litter—hauled in from a local whiskey-barrel factory.

On a hot day, we could cool off in a bin of shelled corn.

We also got to hang around as Grandpa did the necessary building and repair jobs between regular chores. Fence repair was a constant. For that, he carried the necessary tools, including an early multitool, a hammer-like object that was also a pair of pliers and a pry bar all in one, in a metal box attached to the rear mudguard of his tractor. The repairs had to be quick and effective to keep the livestock penned. Extending the concrete pad for the hog shed involved the backbreaking labor of filling a borrowed concrete mixer with sand, gravel, and Portland cement. Poured incrementally over many days, the pad was neither exactly square nor perfectly level, but it served. There just wasn't time for architectural perfection; there were animals to feed and fields to cultivate. I came to admire the solid, no-frills skills required in farming.

One major project took place before my time. The farm centered on a 1910-vintage barn. As the years wore on, the barn started to lean away from its brown-glazed-brick silo. My grandfather hired a carpenter who was a genius with large wooden structures, though not highly skilled at interior work. He spent a couple of days prepping the barn, stringing pulleys and ropes throughout the haymow. He prepared splints and cross

braces, pounding the nails partway in so they would be ready for quick installation. He pounded out some of the pegs locking hand-cut mortise-and-tenon joints. Last of all, he ran several ropes out of the entrance and had Grandpa back his orange Allis-Chalmers tractor up to the barn.

With the ropes tied to the hitch, Grandpa wrestled the tractor into gear and eased forward. With great creaking and groaning, the barn began to right itself, easing back into its original shape, old joints finding their way back home. While the tractor held the tension, the carpenter scrambled over the interior, fastening splits and braces in place. The result: a barn renewed.

I hope you don't have to tackle something that massive on your backyard homestead, but the story always reminds me that with farm structures, perfection is not the goal. What we aim for is solid, utilitarian effectiveness. That makes backyard-homestead projects a great way for beginners to learn carpentry and other how-to skills. A wall slightly out of plumb or a rip cut that wanders a bit aren't that important as long as the structure you are building stands firm and keeps out the weather. After all, chickens are not bothered if a coop door doesn't fit perfectly; goats don't mind if a fence post leans a bit.

This new edition includes 10 brand-new projects, several of which include ideas that have come into their own since the book first published more than 10 years ago. For example, the keyhole garden (page 30), pioneered in Africa, was little known in the wider world. Wind turbines (pages 222-223) are more efficient and reliable than ever. And then there are projects that are just flat-out good ideas—things like the stacked tray vertical planter (pages 40-43). Who knows what the next 10 years will bring?

About the Projects in This Book

Because we know that your time is valuable and your skill level may be only average (or a bit above), the projects in this book are designed with simplicity in mind. If we introduce a somewhat challenging technique—like plunge cuts to make the openings in the coop-and-run project beginning on page 124—it is because, in the long run, it is the simplest, quickest way to get the job done.

A few chapters necessarily focus on what is involved in building the project rather than step-by-step instructions in exactly how to build it. Aquaponics is one example. Whole books and manuals are available on the topic; our chapter equips you with a fundamental understanding of the subject so that you will have a leg up should you want to pursue it.

We also designed these projects with your budget in mind. Each makes the most out of basic materials. There are plenty of gorgeous chicken houses out there, for example (some that would make a decent little cottages for human habitation, complete with clapboard siding, window boxes, Dutch doors, and cupolas), but we went a more utilitarian route, leaning heavily on exterior plywood and simple detailing.

And we paid attention to the human factor—making the finished project convenient to use. Feeding, freshening the water, mucking out, changing litter, egg gathering—all will, we hope, happen more often and be done better because the structure is designed with easy access in mind. You will also find help on how to expand or contract the projects to suit your needs.

Getting Started

Here are some friendly-neighbor-over-the-fence tips that may help as you plunge into a project:

■ Make the exterior screw your default fastener. Predrilling and driving screws takes a bit longer than nailing, but screws hold much better, and you can back them out if you make a mistake.

■ If a circular saw seems too much machine for you to handle, use a saber saw instead. With a square or other straightedge as a guide, it yields a neat, true cut.

■ Measure twice, and cut once.

■ Support your work when sawing so that the material will not bind when cut.

■ Set up a clutter-free work area. It will save time in the long run, produce better work, and keep you safe.

■ Never cut all of your components in advance in kit-like fashion. Instead, work from your project to make sure the measurements for the new piece suit what you have done thus far. Why? Dimensional lumber may vary in size. In addition, small variations as you cut will compound themselves, affecting other areas of the project.

■ Wear eye, ear, and respiratory protection.

■ Gloves make heavy chores seem to go easier because you are not concerned about splinters and abrasions.

■ Improvise! Backyard homesteading is a great laboratory for trying new ideas. If something does not quite work as planned, you can always undo it . . . that is part of the fun.—*David Toht*

CHAPTER 1
Garden Structures

THE HEART OF ANY BACKYARD HOMESTEAD is its garden. While the jury is out on whether the household budget benefits from keeping chickens or goats, there is no doubt that a garden does not just provide you with a bounty of fresh vegetables—it saves you money. (Keep a record of the produce you harvest, adding up what you would pay for that produce at the farmer's market or grocery store, and you will be amazed.) Add to that the opportunity to grow otherwise unobtainable heritage varieties, and you will understand why a backyard homesteader looks first to upgrading the garden.

This chapter offers projects that suit large and small spaces, country and urban. We have tried to cover the basics, like the simple raised bed that leads off this chapter, as well as ideas that are a little out of the norm. The emphasis is on projects that will make the most of your limited space, including a couple of arbors for putting overhead space to good use.

The chapter also deals with that ultimate overhead space, your rooftop. If you are an urban gardener with a building-shaded strip of backyard—or no backyard at all—the roof might be your only recourse. Rooftop gardening means coming up with container systems that are light in weight and easy to hydrate: keeping your plants watered in the face of intense sun, radiating heat from the building, and wind can be a challenge.

The chapter also includes a simple grow-light stand to help you get the jump on the season with plants of a type and nourished in a way that the nursery or home center can't supply. You will discover that for a modest investment you can produce hundreds of starts and save greatly on store-bought starts, which typically cost $4 to $8.

Climbers

Poles, cages, and trellises are the simplest structures you need to add to your garden. Come harvest time, they carry a heavy load and, as anyone who has had a robust tomato plant bring down a store-bought wire cage knows, aren't always up to the job. Here are just a few better ideas:

A bent section of hog fence (those handy grids made of galvanized 1/4-in. (6mm) steel rod), above left, stands on its own to support tomatoes or other heavy crops.

If your beans can spread upward, there is less of a chance that you will overlook ripe pods. This simple rig made of 2x2s and twine, above right, gives runners room for growth.

This wooden trellis, right, lets you walk underneath for harvesting. Its simple construction lets you dismantle it for storage: the crosspieces (inset) slide off to release the grids.

A good, strong limb with splaying branches, far right, is tailor-made for supporting a heavy harvest. A bonus is the attractive way it stands in as a trunk for the plant.

12 Building a Wooden Raised Bed

19 Constructing a Concrete-Block Raised Bed

27 Irrigating a Rooftop Raised Bed

30 Building a Keyhole Garden

36 Making a Bottom-Watered Container Garden

40 Making a Vertical Planter

44 Accessible Gardens and Paths

46 Building an Inclined Planter

49 Building an Arbor

60 Adding a Trellised Arbor

67 Building an Cucumber Trellis

68 Installing a Tool Storage Rack

71 Making a Grow-Light Stand

75 Making Soil Blocks

Building a Wooden Raised Bed

It's no surprise that the raised bed is a fixture of most backyard farms. At 4 feet (1.2m) wide, the generally accepted width, it lets you easily reach into the bed for planting, cultivating, and harvesting without compacting the soil. It does so at a convenient height and limits the incursion of creeping weeds. The length is up to you, though boards longer than 16 feet are expensive and hard to find.

There is also the notion that the soil in that box is yours to take care of. Sequestered as it is, the soil is not going to run off into the pathway. You can nurture it with plenty of fresh compost and all the amendments it needs. You can dig deeply and plant with greater density than with row crops.

This project shows how to stack planks for extra depth. It uses 2x8s, but you could use the same technique to stack two 2x12s to yield a bed nearly 2 feet (61cm) tall—a handy working height. Should you lengthen the bed, be sure to add stakes or pound in rebar alongside the planks to keep them from bowing. The hydrostatic pressure of soil loaded with water can be powerful.

Tools	Materials
Shovel	2x8, 2x10, or 2x12
Rake	pressure-treated
Wheelbarrow	lumber
Measuring tape	2x4 pressure-treated
Level	lumber
Circular saw	2 1/2-in. (6.4cm)
Cordless drill-driver,	deck screws
bits	3-in. (7.6cm)
Framing square	deck screws
Speed square	
Spring clamps	
Baby sledge	
Sawhorses	

A raised bed makes planting easy by raising the work surface and guards the soil from compaction because the gardeners walk on the path, not the garden—both helpful things when you want to get the kids involved.

Raised Bed, Exploded View

Stakes have the dual purpose of strengthening the corner joints and anchoring the raised bed in position. In addition, they help reinforce the walls of the bed. The 2x4 cap is optional, but it helps stiffen the walls and is a comfy place to sit while weeding.

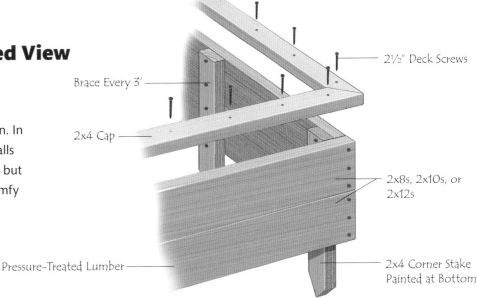

Brace Every 3'

2x4 Cap

2½" Deck Screws

2x8s, 2x10s, or 2x12s

Pressure-Treated Lumber

2x4 Corner Stake Painted at Bottom

Is Pressure-Treated Lumber Safe?

Many gardeners shun the use of any pressure-treated (PT) lumber, fearing that their produce will pick up harmful chemicals. The good news is that chemicals used in the treatment have changed over the years.

Prior to 2003, chromated copper arsenic (CCA) was the prime ingredient for protecting wood from rot. It was more than up to the job but toxic to people and animals. The Environmental Protection Agency (EPA) banned it.

Suppliers turned to two alternative preservatives— alkaline copper quat (ACQ) and copper azole (CA-B). Both contain copper and a fungicide but no arsenic. Does that make PT lumber safer?

Dr. Sally Brown, Research Associate Professor at the University of Washington, makes a study of soil health and how toxins are transmitted to living things. She emphasizes that leaching, if any, is extremely minimal— something on the order of 2 millimeters from the source of the toxin—about the width of a lower case E on this page. "Copper is a necessary nutrient," she emphasizes. "And our bodies are really good at getting rid of extra copper." She has studied plants growing on land covered with copper mine tailings and found little absorption of copper.

She also points out that even if plants next to pressure-treated wood pick up copper, it wouldn't pass beyond the root system. "If this is something that concerns you, don't plant potatoes or root crops along the edge of the raised bed."

That said, if you are commercially growing certified organic produce, be aware that the OEFFA (Ohio Ecological Food and Farm Association) allows only grandfathered PT lumber on structures already on a property but no new construction with PT. Though free of arsenic, the newer copper-based preservatives are not acceptable for soils growing certified organic produce.

The choice is up to you, but bear in mind that unprotected wood will eventually succumb to rot if untreated. That potentially puts to waste your time, effort, and materials.

Pressure-treated (PT) lumber comes in two colors, tan (left) and the greenish hue shown in the following project. The type of lumber ranges from 2-bys to fence planks and even plywood. The parallel incisions let the preservative penetrate the wood.

Building a Wooden Raised Bed

1

Smooth and level the site, clearing an area a bit larger than needed for the raised bed. Remove any obstructing roots or rocks.

The Stacked-Beam Approach

By stacking 4x6s Lincoln Log fashion, you can achieve a substantial bed that will have no problem holding its own without bowing. Side-bolting with 8-inch gutter screws holds this bed together.

2

Cut the side and end boards to length. For an absolutely square corner, you may need to trim the ends. For a straight cut, use a speed square as a guide or clamp a framing square in place (inset). If you are cutting long boards in half, support each half well.

3

Fasten the corners of the 2x8s, 2x10s, or 2x12s with 3-in. (7.6cm) deck screws. Drill pilot holes to avoid splitting.

4

Drive a pointed 2x4 stake about 2 ft. (61cm) long at the highest corner, leaving enough sticking up for the second course, plus a bit more that you'll trim off later. Attach boards to the 2x4 using 2½-in. deck screws.

5

Square up the corners using a framing square. With the first stake serving as an anchor, work your way around the box, tapping the boards into square.

6

Pound a stake into each of the remaining three corners. Level each side, and fasten it to the stake with a single 2½-in. (6.4cm) deck screw. Work your way around to the starting point. If you are satisfied that it is level, finish fastening to each stake.

7

Cut and add the second course of boards. For a tight joint, fasten the corners first (inset); then attach them to the corner stakes.

Building a Wooden Raised Bed (cont'd)

8

Add 2x4 stakes midway along the length of the board to avoid bowing when it is filled with soil. If you leave a bit protruding above the board, you can trim it later, as with the corner stakes.

9

Fasten the stakes using 2½-in. (6.4cm) deck screws. In some cases you may have to use a clamp to draw the stake up to the planks.

10

Trim all of the 2x4 stakes flush with the sides using a handsaw. If you use a circular saw, cut from the stake side with the blade extended just enough to cut through the 2x4.

11

Make a 45-degree angle cut on the ends of the 2x4 caps using a speed square as a guide. Miter one end at a time; position the miter; then mark for the cut at the opposite end.

12

Fasten the caps using 3-in. (7.6cm) deck screws. Fasten the ends first to fit the miters; later, you will fasten every 12–16 in. (30.5–40.6cm) along the run of the cap.

13

Refine the miter joint if necessary by temporarily fastening the mitered 2x4 caps. Then run a handsaw or circular saw (inset) through the joint so that the blade cuts both sides of the miter joint.

14

Fasten the miter joint from the side using a 3-in. (7.6cm) deck screw to draw up and hold the joint. Drill a pilot hole first.

The Quickest Raised Bed

For a fast solution ideal for putting a portion of a driveway or patio to productive use, try a straw-bale garden. Buy several bales of straw (make sure you specify straw, not hay), and push them as close together as possible. Thoroughly soak the bales with water over several days. Chop out the straw where you want your plants, and add topsoil. A straw-bale garden lasts a couple of years before beginning to rot down into great compost fodder.

Building a Wooden Raised Bed (cont'd)

15

Fasten the 2x4s to the raised bed using a 3-in. (7.6cm) deck screw every 12–16 in. (30.5–40.6cm) or so.

Use Exterior Fasteners

An outdoor project is only as good as the fasteners that hold it together, so be sure to use fasteners intended to withstand prolonged contact with moisture. Exterior screws are your best bet because you can easily back them out if you make a mistake—and they hold like crazy. Deck screws are the best choice because their epoxy coating makes them easy to drive. Galvanized (zinc-coated) screws bear up fine, but are more difficult to drive. Where nails are called for, use only hot-dipped, not electro-plated, galvanized fasteners.

16

Fork up the ground thoroughly so that earthworms and other beneficial creatures will have access to the bed.

17

Fill the raised bed with well-amended soil. A few days later, top it off to make up for the inevitable settling of the soil.

Constructing a Concrete-Block Raised Bed

If using pressure-treated (PT) wood concerns you or if you are looking for a truly permanent raised bed, consider using 4-inch (10.2cm)-wide concrete blocks. (See page 13 for more on PT lumber.) By stacking the blocks and filling the voids with concrete, you can construct a handsome bed that will give decades of service.

The greatest advantage to using concrete blocks is that you can build high for a comfortable working arrangement. In addition, blocks and the gravel footing beneath them do a better-than-average job of protecting from creeping weeds—at a cost comparable with lumber.

Prepare for this project by leveling an area about 2 ft. (61cm) wider and longer than your proposed raised bed. The bed shown in this project is 4 × 12 feet (1.2 × 3.7m). The two courses of block add up to a bit more than 16 inches (40.6cm) once you have added the concrete cap—something unachievable with the 2×8s typically used for raised beds. To gain more depth, dig an area about 6 in. inside of the blocks and frame it using 1x4s or 1x6s. Build the frame so that it can be lifted out and used again. (See Step 14.) Lay down a bed of gravel—⅝-inch (1.6cm) crushed gravel was used in this project. In cold regions, include a trench 6–12 inches (15.2–30.5cm) deep and fill it with gravel for drainage. If you plan to add drip irrigation, run the supply lines before laying block.

Tools	Materials
Shovel	⅝-in. (1.6cm) crushed gravel
Rake	
Fork	44 4 x 8 x 16-in.
Tamper	(10.2 x 20.3 x 40.6cm)
Measuring tape	blocks (4 x 12-ft. [1.2 x
Carpenter's level	3.7m] bed)
Torpedo level	4 4 x 4 x 16-in. blocks
Speed square or	(4 x 12-ft. [1.2 x 3.7m]
framing square	bed)
Mason's line and stakes	Construction adhesive
Circular saw	2 2-ft. (61cm) rebars
Work gloves	2 8-ft. (2.4m) 1x4s
Eye and ear protection	4 12-ft. (3.7m) 1x4s
Baby sledge	
Mallet	
Cordless drill-driver	
Mixing tub	
Large caulking gun	
Trowel	
Edger	

A raised bed made of concrete block provides a lifetime of convenience and productivity. This project is made without mortared joints—ever the nemesis of do-it-yourselfers. Instead, blocks are joined using construction adhesive and concrete fill.

The 1½-in. (3.8cm) concrete cap finishes off the block and adds structural strength. Formed using 1x4s, it is easily finished using a trowel and edger.

Block Options

A standard concrete block is 8 x 8 x 16 (20.3 x 20.3 x 40.6cm)—7⅝ x 7⅝ x 15⅝ (19.4 x 19.4 x 39.7cm) to be exact. A block 6 or even 4 (15.2 or even 10.2cm) wide is more than up to the job. Half blocks are needed because of the Lincoln Log overlap at the corners. Some 6- and 8-inch-wide (15.2 and 20.3cm) blocks come with three voids in them.

Constructing a Concrete-Block Raised Bed

1

Prepare the site by leveling the area and digging down a few inches to deepen the bed. Install a temporary 1x4 frame as shown, and level it. With the frame in place, spread gravel for the paths and a footing for the block. Using stakes or rebar, reinforce the frame.

2

Check the frame for level. Check between frames, too; use a straight plank to extend the level's reach if necessary. You need at least 3 ft. (91.4cm) for a pathway between the beds—enough room for your wheelbarrow or garden cart. Using the frame as a guide, rake the gravel level.

3

Screed the gravel using a 2x6. Place the plank on the frame, and push it to level out the gravel.

Constructing a Concrete-Block Raised Bed (cont'd)

How to Set a Block

Take the time to set each first course precisely, leveling in two directions with a torpedo level. Get them right, and the second course will be a breeze. Here's how:

1. Level across the width of the block. Rock the block to settle it in. Add additional gravel as needed.

2. Level the length of the block. Make sure it is tight against the adjacent block.

3. Use a mallet to make adjustments. A hard rubber or wooden mallet will persuade the block without chipping it.

4

Tamp around the frame, filling in with extra gravel as needed. Tamping will compact the gravel a bit below the frame edge. Keep it as smooth and even as possible.

5

Begin laying the blocks, leveling and adjusting as you go. (See "How to Set a Block," left.) Begin at one of the ends, aligning the blocks carefully.

6

Knock out any excess concrete in the second-course blocks. A little buildup occurs in manufacturing, and you must tap it away using a steel stake or piece of rebar so that the concrete you pour will fill the voids completely.

7

Build up one end by applying construction adhesive before setting the second-course blocks in place. Note that you will need a half block (inset). Use a masonry level or carpenter's level on a straight plank to confirm that both sides are even. Stagger every joint.

8

Complete both ends of the raised bed, seating each block in the adhesive with a few taps of the mallet. The second course will be about ½ in. (1.3cm) longer—a result of the blocks being designed for mortar joints.

9

Finish the first course of both long sides, using masonry string as a guide. If all goes well, you will need to just ease in the final block.

Constructing a Concrete-Block Raised Bed (cont'd)

10

Add the second course, using construction adhesive to seal the blocks in place. The second course goes quickly because you don't need to level the blocks.

11

A strap with a ratchet is ideal for cinching things up and stabilizes the block as you add concrete. Use a length of rope as an alternative. Now is the time for dealing with any gaps between blocks.

12

Pound in a 2-ft.-long (61cm) piece of rebar to stabilize the midsection of both long sides. Do not add rebar to the corners—the bars will clog the voids, keeping them from being filled with concrete.

13 Fill the voids in the blocks with concrete, mixed with enough extra water so that it works easily. Stop an inch short of completely filling each void—the concrete cap will make up the difference. Use a piece of rebar (inset) to work in the concrete.

14 While the concrete is setting up, remove the frame (inset) and thoroughly fork up the ground. Add soil, and spread it evenly.

15 Cover the ground with a sheet of commercial-grade weed block, cutting it large enough so that it covers the interior wall by about 6 in. (15.2cm).

Constructing a Concrete-Block Raised Bed (cont'd)

16

Make a form for the cap by cutting 1x4s to fit the interior and exterior edge of the bed. Use a scrap of 2x4 as a guide for positioning them as you clamp them in place (inset).

Starting Right

To get your raised bed off to the best possible start, mix compost, chopped peat moss, and sand, as well as amendments (as needed) such as kelp meal, bone meal, colloidal phosphate, lime, and greensand. This light and absorbent mix provides a great medium for your plants and will be easy to fork over next season. Because of the depth of the bed, plants will send their roots down deeply. Some gardeners plant beets and onions in double rows 4 inches apart every 12 inches (shown). The soil medium will compress over time, so plan on adding to it each season.

17

Shovel concrete into the form, and smooth it using a masonry trowel. Tap the forms to eliminate voids.

18

Round the edges of the cap by working an edger along the inside of the form. Let the cap cure overnight before removing the form. Give the entire project a few days to set up before filling the raised bed with soil.

Irrigating a Rooftop Raised Bed

Any rooftop gardener will tell you that keeping raised beds adequately hydrated is a constant battle. The sun, heat radiating from the roof, and wind conspire to evaporate water. The result is soil much thirstier than that of a ground-level garden.

This inexpensive scheme not only puts water where it can do the most good but reduces the amount, and therefore the weight, of soil required—an important consideration for rooftop gardeners. It does so with buried drain tubing, easily filled thanks to a cutoff plastic bottle that funnels water to the tubing. A weep hole prevents overwatering. Weed block keeps the soil from clogging the tubing. With plenty of water at root level, plants flourish.

Brooklyn apartment dweller Adam Matthews fills his raised bed once a week, looking down the plastic bottle to check the water level in the bed. "The dirt basically functions like a sponge," he says. "The places where the landscaping fabric dips down between the drain tubes is where the dirt sucks up the water."

Tools	Materials
Utility knife	Heavy-duty plastic
Shears	4-in. (102mm) corrugated
Stapler	and perforated drain
Drill-driver and	tubing
½-in. (12mm) bit	Weed-block fabric
	Plastic bottle
	½-in. (12mm) tubing

Despite the evaporation from heat and wind, this rooftop raised bed grows water-intensive crops like kale and cabbage with ease, thanks to buried drain tubing that slowly releases water to plant roots.

Irrigated Raised Bed, Cross Section

As one drainpipe is filled, water migrates to the others. Weed block keeps the soil from clogging the tubes. A short length of plastic tubing set in the side of the bed signals overfilling.

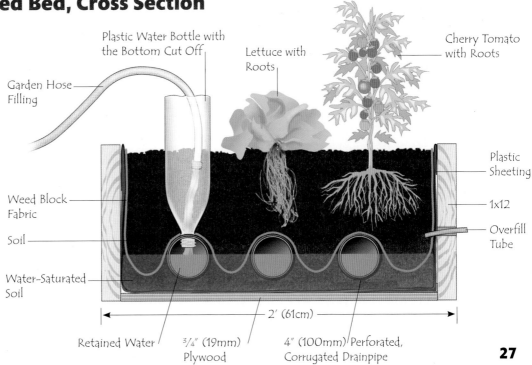

Plastic Water Bottle with the Bottom Cut Off

Garden Hose Filling

Lettuce with Roots

Cherry Tomato with Roots

Plastic Sheeting

Weed Block Fabric

Soil

1x12

Overfill Tube

Water-Saturated Soil

Retained Water

¾" (19mm) Plywood

4" (100mm) Perforated, Corrugated Drainpipe

2' (61cm)

Irrigating a Rooftop Raised Bed

1

Spread plastic sheeting and a 1-in. (2.5cm) layer of soil on the floor of the bed to cushion the plastic and avoid punctures. Check for rocks, nails, and splinters—anything that might pierce the plastic.

2

Lay in another sheet of thick plastic and run it up the sides of the bed. Staple only along the top edge of the bed.

3

Add perforated drainpipes to the bottom of the bed. Using a utility knife, cut them so they fit tightly against the ends of the bed.

4

Cover the pipes with weed-block fabric, tucking the fabric between the tubes. Staple the fabric to the walls, and trim away any excess. At the end of one pipe, use the utility knife to cut an "X" through the fabric and pipe wall. Cut off the bottom of a plastic bottle. Push it upside down into the hole.

5

Test-fill the bed, and note the point at which the pipes seem to fill completely. Check for any leaks.

Keep It Light

Rooftops can bear only so much. Don't tempt fate by using plain backyard soil. Instead, lighten soil or good compost with absorbent materials like pumice, coconut fiber, peat moss, perlite, expanded slate, or vermiculite. It's possible to blend these materials with as little as 25 percent soil or compost to come up with suitable rooftop soil.

6

Bore a hole in the side of the bed, and insert ½-in. (12mm) tubing as an overflow.

7

Fill the bed with your soil medium of choice. Plant sets and seeds as you normally would. Be careful not to pierce the plastic when adding poles, stakes, or cages.

Building a Keyhole Garden

Form and function come together in the keyhole garden. Designed for ease of gardening and water conservation, a keyhole garden can be a good-looking, highly effective addition to your backyard homestead.

Designed for intensive gardening in arid regions, a keyhole garden combines several attributes. First, it is a raised bed, easily accessible, and because it is not stepped on, it has little of the soil compaction of a typical garden. Second, it has a ground-level layer of stones and twigs that store excess moisture for the dry season. Finally, accessed by a "keyhole" pathway, it incorporates a compost bin to leech goodness into the surrounding garden.

Keyhole Garden Origins

Based on a design that originated in Zimbabwe, keyhole gardens were adapted in Lesotho, South Africa to meet a special need. Lesotho had an exceptionally high number of HIV/AIDS sufferers who couldn't manage the usual digging, planting, and weeding required by a conventional garden. The keyhole scheme put everything within reach, even if the gardener was seated. In addition, it conserved precious rainwater. The method proved so productive they caught on as kitchen gardens for the broader population and soon spread throughout the region.

Getting Started

A keyhole garden can be made of stone, brick, cinder blocks, pavers, wattle, or wood—anything strong enough to contain the soil. This project uses pressure-treated (PT) decking and 2x4s. (For details on the safety of PT lumber, see page 13.) Exact dimensions are provided to make the structure shown. If you choose to expand it, don't let the reach required for weeding and harvesting exceed your grasp. The best orientation for the keyhole access opening is to the north. That way the effect of the compost bin shadow is minimized.

Tools	Materials
Eye protection	12 12-ft. (3.7m) 5/4x6
Ear protection	PT decking
Tape measure	8 12-ft. (3.7m) PT 2 x 4
Marker	2 100-ft. (30.5m) rolls of
Chop saw, miter saw,	strapping
or circular saw	500 self-tapping ¾-in.
Jigsaw	(19mm) lath screws
Extension cord	2 4-ft. (1.2m) steel posts
Drill/driver	3 3-ft. (91.4cm) steel posts
Wheelbarrow	½-in. (12mm) staples
Stake and line	50-ft. (15.2m) roll of
Shovel	weed-block
Clamps	landscaping fabric
Speed square	Corrugated cardboard
Tin snips	Rocks
Baby sledge	Sticks
Stapler	String

Best Vegetables to Plant

Deep reaching plants like onions, carrots, radishes, kale, lettuce, spinach, and turnips take better advantage of a keyhole setup than plants with spreading root systems like squash and tomatoes. Because a keyhole garden is deep, plants can be spaced closer together than recommended on the seed packet. Often the distance can be reduced 50 percent or more.

Keyhole Garden Plan View

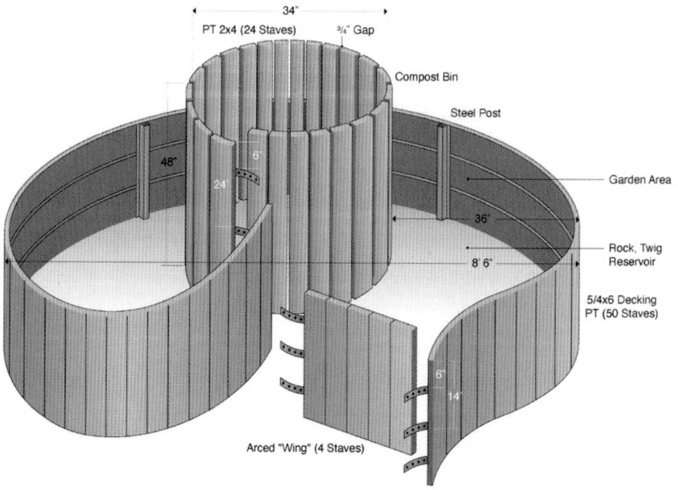

34"
PT 2x4 (24 Staves)   ¾" Gap
Compost Bin
Steel Post
48"
6"
24"
Garden Area
36"
8' 6"
Rock, Twig Reservoir
5/4x6 Decking PT (50 Staves)
6"
14"
Arced "Wing" (4 Staves)

Building a Keyhole Garden

1

Clear a space for your garden. Then, center a stake and string equal in length to the radius of your keyhole garden—in this case, 52 in. (132.1cm). Strike a circle in the dirt.

2

Cut 60, 28-in. (71.1cm)-long pieces of 5/4 x 6 pressure-treated decking. Use a simple jig as shown to mark them.

3

Decking is heavy, so fabricate the perimeter in small sections. Line up the ends of six or eight pieces along a straight edge. Attach strapping 3 in. (7.6cm) from the bottom edge and 3 in. (7.6cm) from the top edge using self-tapping ¾-in. (19mm) lath screws. Leave 5 in. (12.7cm) of strapping protruding beyond one side of each section to use later to link sections together.

4

Prefab the two "wing" sections that will sweep upward toward the compost bin. Rough cut four pieces ranging in length of about 30, 33, 35, and 38 in. (76.2, 83.8, 88.9, and 96.5cm). Fasten two straps, one 6 in.(15.2cm) from the bottom edge and the other 18 in. (45.7cm) above it. Using a thin scrap of wood and some string, make an arced bow. (This need not be exact—just something that pleases your eye.) Strike a line along the bow, starting at 28 in. (71.1cm) from the bottom at one end and ending at 38 in.(96.5cm) at the other.

5

Clamp the wing section and cut the arc using a jigsaw. Remember to start the saw before making contact with the wood. Follow the line as best you can. Aim for a smooth arc overall.

6

Make mirror image wings to assure that the strapping will be on the inside of the keyhole garden.

7

Cut the 2x4s for the compost bin using a chop saw or circular saw. Cut 24, 48-in. (121.9cm) pieces. As with the perimeter staves, a simple jig assures uniformity.

8

Prefab sections of the compost bin. Like pressure-treated decking, PT 2x4s are heavy, so prefab sections of four to six staves. Mark each stave 6 in. (15.2cm) from the top and bottom, and midway at 24 in. (61cm), for positioning the galvanized strapping. Line up the ends of the pieces along a straight edge. Using a 1x2 spacer, attach the strapping with lath screws. Leave 5 in. (12.7cm) of strapping extending from one side of each section.

Building a Keyhole Garden

9

Assemble the compost bin overlapping the excess 5 in. (12.7cm) of strapping to attach the sections together. Use a 1x (¾-in. [19mm]) spacer when joining two sections.

10

Pound in the steel posts to stabilize the compost bin. Two 4-ft. (1.2m) posts are enough to keep it from being tippy until the soil is in place.

11

Attach the perimeter wings to the compost pile using lath screws. Position the wings to make a pleasing curve toward the outer perimeter.

12

Attach the perimeter sections, beginning at a wing. As you choose a prefab section, be sure to always keep the neatest edge upward. Work all the way around to the other wing, forming a smooth circle. Pound in four 3-ft. (91.4cm) posts to stabilize the perimeter.

Add weed block to keep soil from escaping between the uprights. Stapling weed block fabric to the inside of the perimeter gets the job done until the soil is added.

Line the garden area with cardboard. Add branches and twigs to create the reservoir that will hold moisture during dry spells. Shovel in the best soil you can get your hands on, well-blended with mature compost.

Plant the garden, and start loading the compost bin. Because of the soil depth, you should be able to cut the recommended plant spacing in half. The soil will settle during the first year, requiring a top-up when you dig again the next year.

Making a Bottom-Watered Container Garden

To counter the effects of dehydrating wind and sun, this rooftop garden box keeps soil wet by wicking up moisture from a water-filled compartment.

Bottom-watered containers are famous for producing a bounty of vegetables in a small space. You can purchase them ready-made, but building one yourself is easy to do.

You may use any of a variety of containers, including 5-gallon buckets or large plastic pots, to make one. This project uses an 18-gallon plastic bin with a lid, big enough for a couple of tomato or pepper plants or up to six smaller plants. Surrounding the bin with a wooden box makes it look great and provides a base for a simple trellis.

With a fill tube for topping up the water supply, this bottom-watered container tolerates the intense sun and dehydrating wind of a Chicago rooftop. A decorative box of inexpensive cedar fence boards and a 2x2 trellis complete the package.

Tools	Materials		
Utility knife Felt-tip marker Drill-driver with ½- and ⅜-in. (12 and 10mm) bits Garden hose	18-gal. plastic storage bin, with lid 3 ft. (91.4cm) or so of 4-in. (102mm) PVC pipe (either regular or perforated)	2 ft. (61cm) of 1-in. (25mm) PVC pipe Adhesive good for joining plastic to plastic Large bag potting mix (not potting or garden soil)	1 lb. of dolomite (lime with calcium and magnesium) 2–3 cups of granular fertilizer

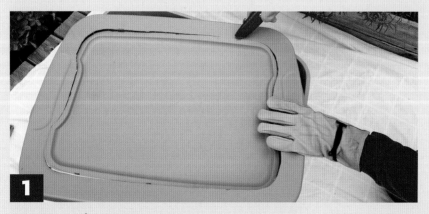

1

Cut an aeration screen in the bin's lid. Measure the inside of the bin in both directions (inset), 6 to 7 in. (15.2 to 17.8cm) up from the bottom. Place the lid on a flat surface; mark it for the cut using a felt-tip marker; and use a utility knife with a new blade to cut it. The cut need not be exact; a gap of up to ⅜ in. (9.5mm) around the perimeter is fine.

Making a Bottom-Watered Container Garden

2

Cut five lengths of 4-in. (102mm) PVC pipe to 6 or 7 in. (15.2 or 17.8cm) Use a power miter saw or hand miter box, or cut carefully with a power saw so that the ends are straight and all of the pipes are the same length. Also cut a length of 1-in. (25mm) PVC to about 2 ft. (61cm) Using a drill equipped with a ½-in. (12mm) bit, drill a grid of 10 or so holes in each of the 4-in. (102mm) pipes (inset). Also drill six or so holes in the bottom 6 in. (15.2cm) of the 1-in. (25mm) pipe.

5

Drill a grid of fairly evenly spaced ⅜-in. (10mm) holes in the aeration screen. This will allow water to seep up into the soil but not allow the soil to drop into the water. Use a utility knife to scrape away any burrs from the holes.

3

Place the pipes on the aeration screen so that they are evenly spaced. (They support the screen and keep it from sagging when the soil is added.) Mark their positions.

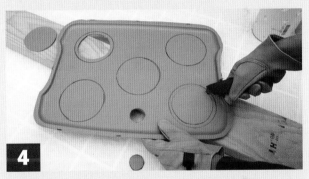

4

Cut holes in two of the marked 4-in. (102mm) circles using a utility knife. Make sure the holes are about ⅜ in. (9.5mm) inside the marked lines so that the pipe can support the screen. Also cut the hole for the 1-in. (25mm) pipe.

6

Attach the PVC pipes to the screen using an adhesive such as cyanoacrylate or two-part epoxy made for joining plastic to plastic. Apply the adhesive to the pipe ends; position them; and place a weight on top. Allow the adhesive to set. Alternately, use twist ties or lengths of wire to hold them in position.

Making a Bottom-Watered Container Garden (cont'd)

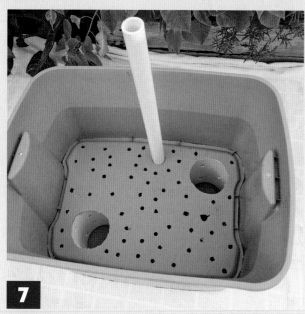

7

Place the aeration screen into the bin, and slip the 1-in. (25mm) fill pipe into its hole.

8

Measure up to a point about ½ in. (12.7mm) below the top of the aeration screen. Drill a ½-in. (13mm) hole at this point. This will allow water to drain out and not soak the soil above the screen.

9

Pour potting mix (not potting soil or garden soil) into the two drainpipes; then fill the bin with about 4 in. (10.2cm) of mix. Sprinkle water onto the mix until it is thoroughly dampened.

10

Add 4 in. of potting mix while sprinkling so that all of the mix is damp. Continue until you are about 6 in. (15.2cm) from the top of the bin. Sprinkle 1 lb. of dolomite evenly over the surface, and then add the final 4 in. (10.2cm) of potting mix.

11

Make a 2-in.-deep (5.1cm) trough across the length of the bin. Plan the location of the trough so that it is a few inches away from where you will place the plants. Pour in 2–3 cups of granular fertilizer, and cover the trough.

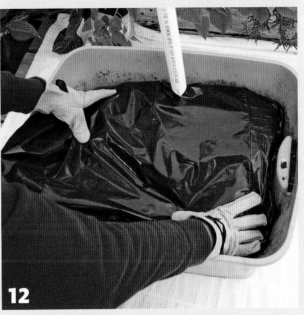

12

Cut a piece of heavy-duty plastic 3 in. (7.6cm) larger than the top of the bin. Place it on top, and tuck it in on all sides. Replace the remainder of the bin's lid to help keep the plastic in place. You may want to set small weights at the corners to keep the plastic from blowing away.

13

Cut a small X-shape hole in the plastic; slip the plant into the soil; and fold the plastic back over. You may choose to plant two large plants, such as tomatoes or eggplants, or as many as six smaller plants, such as broccoli, beans, or lettuce.

14

Add water to the container until it drips from the drain hole (Step 8). If you have a simple drip irrigation system, add a drip emitter to the end of a ¼-in. (6.4mm) line, slipping the emitter into the fill tube. Or add water from a hose every few days.

Making a Vertical Planter

Vertical planters are a great way to grow greens and herbs on your balcony or deck. Whether you have a small space or a large one, fresh produce at your fingertips is unbeatable.

If a balcony, deck, or patio is your only spot for raising vegetables, this vertical planter helps make the most of it. Even if you have plenty of space, you might want to build this project as a convenient source of greens and herbs right outside your kitchen door.

The planter is simple to build and can be completed in a few hours. Once you've made a marking guide and cutting jig for the single angle used throughout this project, things will move quickly. To avoid wasting material, cut the longest pieces first. And if you have access to a chop or sliding miter saw—use it. It'll help you really crank out exact pieces quickly.

This project uses flower box liners as planting trays, but you can substitute most anything that offers 5 or 6 in. (12.7 or 15.2cm) of depth and is rigid enough to be loaded with soil. Aluminum or vinyl gutters are alternatives.

Tools	Materials
Ear protection	5 flower box liners or
Eye protection	equivalent, roughly
Work gloves (optional)	36-in. (91.4cm) long,
Measuring tape	8-in. (20.3cm)
Pencil	wide, 5-in. (12.7cm)
Circular saw (or chop or	deep
sliding miter saw)	12 1⅝-in. (41mm)
Bar clamps	Phillips head
Speed square	exterior screws
Cordless drill/driver	24 1½-in. (38mm)
⅛-in., ⅜-in. (3 and 10mm)	Phillips head
drill bits, Phillips bit	exterior screws
Level	
Sanding block	

Vertical Planter Plan

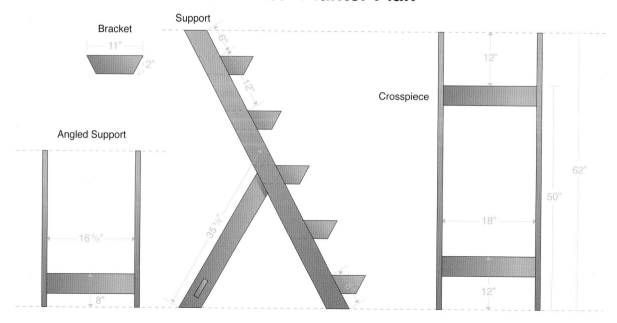

Making a Vertical Planter

1

Make a marking guide for the angle cuts by striking a line 2 in. (5.1cm) from the end of a 1x4. Then, mark a diagonal line from the 2-in. (5.1cm) mark to the end of the board.

2

Use a circular saw to cut the angle. Wear ear and eye protection. Then, make a perpendicular cut about 8 in. (20.3cm) from the angle cut. Now you have a handy marking guide for the whole project.

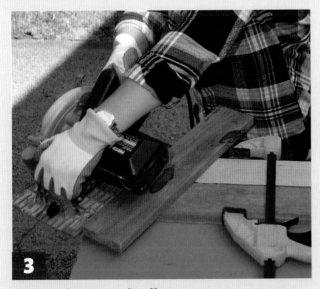

3

Make a simple sawing jig by cutting the angle on a scrap and attaching a guide. Start the cut, and then slide the jig into place and clamp it. To cut an opposite angle, simply flip the board. Cut your first components, the 62-in. (157.5cm) supports. Touch up any rough edges with a sanding block. Do this on all the pieces.

4

Mark the location of the support brackets. Begin by measuring for the first tray bracket 8 inches (20.3cm) from the bottom of the support. Then, mark every 12 in. (30.5cm) for the remaining four brackets. Make an angle cut at the top of the support, 6 in. (15.2cm) above the fifth bracket mark. *Note: The top of each bracket will align with these marks.*

Making a Vertical Planter (cont'd)

5

Cut the brackets. Saw a bracket with two opposing angles as shown. The long side is 11 in. (27.9cm). Use a finished bracket as a template for marking for the rest of the 10 brackets, flopping it over as you go. Cut the remaining brackets.

6

Clamp each bracket in place before attaching it. Double check that you align the top of each bracket with its mark. To avoid splitting the wood, drill pilot holes using an 1/8-in. (3mm) bit. Drill holes for two fasteners per bracket, staggering them as shown—another defense against splits.

7

Fasten each bracket with a 1¼-in. (32mm) exterior screw using a drill/driver equipped with Phillips bits.

8

Fabricate a mirror image support by using the completed support as a guide. Clamp the two together and attach the brackets.

9

Attach the 18-in. (45.7cm)-long crosspieces, positioning them 12 in. (30.5cm) from each end of the supports. Resting each piece on 2x4 scraps centers them nicely. Fasten each with four 1⅝-in. (41mm) exterior screws.

10

Make the angled support by cutting the side pieces so the end up with opposite angles with a long side of 35½-in. (90.2cm). Attach the 16⅜-in. (41.6cm) crosspiece 8 in. (20.3cm) from one end using four 1⅝-in. (41mm) fasteners.

11

Attach the angled support by clamping it so the top of the angle cut is about 34 in. (86.4cm) up the support. Outdoor areas are seldom level, so you might want to make fine adjustments using a level before fastening with four 1¼-in. (32mm) exterior screws.

12

Drill drain holes in your trays to avoid drowning plants when it rains. Four to six ¼-in. (6mm) holes will do it. That done, you are ready to set your trays on the planter, add soil, and plant your vertical garden.

43

Accessible Gardens and Paths

A table garden offers roll-under ease for tending to herbs and greens. The best practices for accessibility are raised beds and quality pathways.

The joy of gardening should be available to all. A raised bed makes that possible for a wheelchair user, especially if the bed is suitably sized and accessible by thoughtfully surfaced pathways. Public access gardens must allow for a wide range of needs and are typically guided by Americans with Disabilities Act (ADA) standards. For home gardens, it is likely that the needs of only one individual is involved, giving you more flexibility in the types of materials you an use. For example, while the ADA is not keen on wood chip paths, your gardener's comfort using such a path can prevail. Here are some factors to bear in mind when planning.

Best Raised Beds

With a table-type raised bed, a wheelchair can be rolled underneath—as long as there is 28–36 in. of clearance. That's a big advantage. However, because a table garden only has about 6 in. of soil depth, the gardening is limited to herbs and greens. Vegetables like radishes, carrots, and peppers need at least 12 in. (30.5cm)—which would make a table bed too high to tend comfortably. And plants like tomatoes, cucumbers, and squash need 18 in. (45.7cm) of soil—way out of reach.

A better bet is a ground-based raised bed of a comfortable height and width. That means a bed 28-34 in. (71.1–86.4cm) high and, if accessible from both sides, about 36 in. (91.4cm) wide. Such a width allows a person using a wheelchair to wheel up to the bed and comfortable reach in to tend a garden.

Pathway Planning

If your space is limited, you can go with a path as narrow as 36 in. (91.4cm). However, that will not leave a wheelchair enough space to be turned to face the bed. The user must turn sideways to work the soil, which is not always a comfortable position.

A better option is a path wide enough to allow the gardener to spin the wheelchair around to face the bed. That requires a path at least 5 ft. (1.5m) wide to allow for a 90-degree spin. Facing the garden is much more comfortable if leaning forward is not a problem.

Best Path Materials

ADA compliant materials like concrete, asphalt, pavers, and wood are ideal but might not be in your budget.

What's more, most of them are not permeable, possibly creating drainage problems. Wheelchairs do best on a surface that is neither too squishy or too bumpy. As mentioned, surfaces like wood chips, pea gravel, sand, and rubber mulch are not considered ADA compliant because wheelchair wheels and crutches cannot gain traction in them. However, ideal surfaces may not be in your budget. Again, the wheelchair user is the final judge on what surface materials are acceptable and affordable.

Fave Paves

Self-binding fine gravels adhere to one another as they are rolled on, making an almost completely smooth surface.

Pavers, set smoothly, are an easy roll.

Bricks and stones can be smooth if closely set and level.

Wood planks, with a 1/4 in. (6mm) between boards for drainage, are easy on wheels.

Bark or mulch, if seasonally refreshed and tamped down, can work for short distances.

Flags, closely set and level, are less than ideal but workable.

By the Numbers

Remember, these are guidelines. Tailor your garden to suit the user!

Best height for a raised bed:	Best width for a raised bed:	Maximum reach into a bed from either side:	Access path width:	Path between beds:
28-34 in. (71.1-86.4cm)	36-40 in. (91.4-101.6cm)	34 in. (86.4cm)	3 ft. (91.4cm)	5 ft. (1.5m), to allow turning a wheelchair to face the garden

Building an Inclined Planter

If your space is limited or you just like the look of high-rise plantings, here's an attractive planter that you can make in 4 to 6 hours. Inclined to give every plant full sunlight, the planter will produce a bounty of herbs or vegetables.

Given its small 3 x 3-ft. (91.4 x 91.4cm) footprint, it can be located on a deck or against the house on a sunny portion of the driveway. Because it is inclined, most of the plants are within easy reach—a big advantage if you want to grab a few leaves of fresh basil. And it is simple to micro-irrigate, a sure way to keep up with the frequent watering that all containers demand.

The pleasing randomness of the pot arrangement on this planter, above right, mimics a hillside, an attractive way to add productive space to your yard.

Tools	Materials
Measuring tape	1 12-ft. (3.7m)
Circular saw	pressure-treated 2x4
Drywall T-square	4 x 3-ft. (1.2m x 91.4cm)
Spring clamps	piece of exterior plywood
Saber saw	(½ to ¾ in. [13 to 19mm] thick)
Cordless drill-driver	1½-in. (38mm) exterior screws
and bits	3-in. (76mm) exterior screws
Scribing compass	A variety of plastic pots

1

Cut a 48-in. length of 2x4, and mark it for the inclined frame by using the corner of your piece of plywood. Align the 2x4 so that its ends just touch the edges of the plywood. Adjust the incline to your liking, and mark the 2x4. Cut it, and use it as a guide for cutting the second piece.

2

Join the angle-cut frame pieces at their midpoints to a 33-in. (83.8cm) crosspiece using 3-in. (76mm) deck screws. Be sure to orient the inclined frame pieces so that the angles match at each end.

Building an Inclined Planter

3 Set the frame, and arrange the pots on the worse side of the plywood. Be careful that the pots along the top and bottom of the planter are far enough from the edge so that they won't protrude so much that they keep the unit from leaning against a wall.

4 Trace around the pots with a marker. This mark is the outer limit for cutting.

5 Draw a cut line by finding the center of the circle and, using a compass, scribing a line equal to the width of the pot less its lip. Double-check your work to make sure that the pot won't fall through once the hole is cut.

6 Bore an access hole, and cut along the scribed line using a saber saw. When done, cut off the 36-in. (91.4cm) section of plywood using a guide as shown above.

Building an Inclined Planter (cont'd)

7

Attach the plywood to the frame. Flip the plywood (the cutout being smoothest on the underside), and position it over the frame. Fasten the plywood to the frame using 1½-in. (38mm) exterior screws every 8 in. (20.3cm).

8

Glue the pots in place using a few dots of silicone caulk or exterior-grade construction adhesive. Be sparing with the adhesive; you may want to remove a damaged pot later.

9

Set the planter on concrete pavers or flagstones to limit water damage. The pavers also make leveling the planter easier.

10

Add a micro-irrigation emitter to each pot if a water supply is nearby. All containers are prone to dry out; an irrigation system on a timer will keep the plants well hydrated.

Building an Arbor

Starved for growing room? An arbor puts to use overhead space too often overlooked. With a modest investment and a weekend or two you can add an attractive arbor to your patio, porch, deck, or walkway and put it right to work. You'll soon have a shaded retreat providing fruit or vegetables.

This arbor was designed to make productive use of the grass strip between the street and sidewalk and be a welcoming feature that would frame the entryway to the house. That meant ramping up the detailing. The overhead beams are notched, a pleasing touch that pulls the elements together. The notches are deep enough that no fasteners are needed. (If you want to skip this detail, hardware ties—Simpson Strong Tie RTB is one—or angled screws will do the job.) Designer/builder Roy McGarrah added benches—and angled detailing to the beams.

Give some thought to siting your arbor. Make sure it gets enough sun to grow your crop of choice and adds shade where you want it. The arbor shown is almost 8 ft. (2.4m) wide (5½ ft. [1.7m] at the bottom), 8 ft. (2.4m) high, and nearly 15 ft. (4.6m) long. The design is flexible, however; the width and length are easily altered to suit your site.

What should you grow on the arbor? Grapes are an ever-popular choice, but blackberries and passion fruit can also work. One caveat: at harvest time, berries and fruit can make a mess of a patio or deck if not picked in time. One alternative is a hop vine; a quick-growing covering that will provide

Joined with notches, the main beams and crossbeams of this arbor need no fasteners. The main structural members are held to the posts with lag screws covered by wood plugs.

fresh hops for the home brewers in your life. Here, kiwi is the vine of choice.

The project shown uses rough-sawn cedar 2x6s and 2x8s, but any 2-by lumber would do as well. The doubled 4x4 posts look great and keep the structure from racking—a strategy for avoiding angled braces. Vines harbor moisture, so guard against rot. We sealed the cedar with linseed oil before beginning construction. With age, the wood will turn a deep gray. Heartwood redwood or cedar is ideal for arbors but increasingly rare and expensive. Pressure-treated lumber is more than up to the job. For help on how to lay out a project like this, see pages 50-59.

This handsome arbor puts the area next to the sidewalk to productive use as a home for kiwi vines. In addition, it is a welcoming feature to the lot (inset).

Arbor Plan

The 8-ft. (2.4m) height of this arbor is about right for any overhead project, but width and length can flex to suit your site. It is almost 8 ft. (2.4m) wide and 15 ft. (4.6m) long—about as massive as you'd want to go. It can readily be downsized a couple of feet or so.

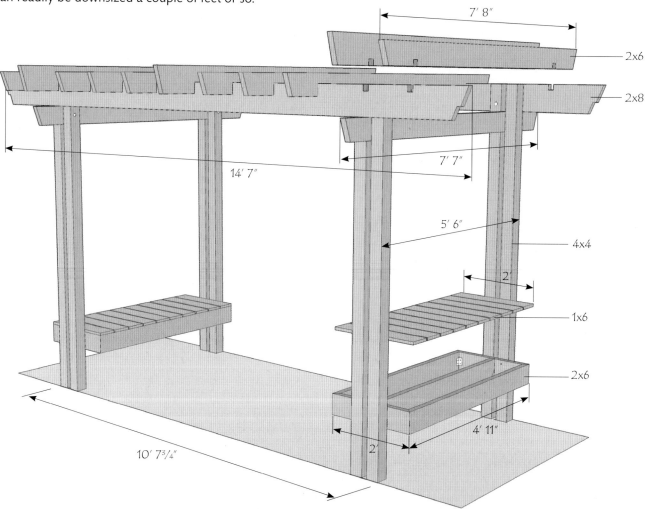

Tools			Materials	
Posthole digger	Saber saw	Sanding block	2 16-ft. (4.9m) 2x8s	3-in. (76mm) deck screws
Shovel	Power miter saw	Large clamps	3 8-ft. (2.4m) 2x8s	8 6-in. (152mm)-long
Wheelbarrow	(optional)	1-in. (25mm)	10 8-ft. (2.4m) 2x6s	⅜-in. (10mm) lag screws
Trowel	Drill-driver with bits	plug-cutter bit	8 10–12-ft.	4d galvanized box
Measuring tape	6-in. (152mm)-long bit	Screwdriver	(3–3.7m) 4x4s	or finishing nails
Speed square	for lag screws	Squeeze clamps	5 12-ft. (3.7m) 1x6s	Raw linseed oil or
Level	Hammer	Socket wrench set or	1 6-ft. (1.8m) 1x6	other sealer
Circular saw	Chisel	adjustable wrench	Scrap 1x2s as braces	

Building an Arbor

Install the posts by digging postholes and setting them in concrete. (See pages 81–83.) For two posts, attach a temporary spacer as shown to keep the gap between the posts consistent. Plumb the posts using a carpenter's level, and brace them using diagonal 1x2s.

Decide the overhang you want for your main beam (24 in. [61cm] in this case) and whether you want an ornamental treatment. Mark it for sawing. Make the angled cuts using a circular saw; then finish them using a saber saw.

Temporarily install the main beam, which is the strongest visual element and establishes the trim height of the posts. Use clamps and blocks to set it in place and level it. Drive 3-in. (76mm) deck screws to temporarily hold it—you'll need to remove it later to trim the posts. Mark the posts for cutting.

Building an Arbor (cont'd)

4

Mark the opposite posts for trimming by setting a straight piece of lumber on the main beam. Check for level, and mark the posts; then do the same for the next pair of posts.

5

Level around the structure by setting the second main beam in place and checking that it is level. Make your marks, and then temporarily fasten the beam in place with 3-in. (76mm) deck screws.

6

Complete the leveling by using a straight piece of 2-by to check the final side. You probably will not come out dead on. If you are ⅛- or ¼-in. (3.2 or 6.4mm) out of level, don't worry about it. If greater than ¼ in., (6.4mm) adjust the second main beam to average the difference.

7

Remove the main beams; mark for the cut on two sides of a post; then clamp a speed square onto the post as a guide for cutting. You will need to measure the base (or shoe) of the circular saw to determine where to place the speed square (inset).

8

Trim the posts by setting the base of the saw on the clamped square. Check that the blade is not touching the wood. Start the saw, and ease it into the cut. Keep the base of the saw flat on the surface of the post.

9

Seal the post tops to protect the open grain of posts from rot. Unless you are using pressure-treated lumber, seal your cuts as you complete them.

10

Mark the main beams for notching by setting them on sawhorses and laying out the crossbeam locations. Confirm that you like the look of the spacing—scraps of wood help you visualize the crossbeams. Mark for each notch using a scrap of 2-by as a guide to depth and width.

11

Adjust your saw's cutting depth using a scrap of lumber. A 1½-in. (3.8cm) notch depth will give the look of a 3-in. (7.6cm) overlap at the intersection of the beams.

Building an Arbor (cont'd)

12 To notch the main beams, line up the ends and clamp the two beams together. Make the first (and last) cuts for the notch using the speed square as a cutting guide. Freehand the rest of the notch by "kerfing," or making repeated side-by-side cuts.

13 Clean up the notches where kerf marks and bits of wood remain. Use a 1-in. (25mm) wood chisel, followed by a sanding block.

14 Make a test fit to check your marking and notching technique. Use a scrap of lumber from one of the crossbeams. Recut as necessary. Check every notch—it is not fun making an adjustment once the beam is in place.

15 Seal the notches. Exposed to the elements as they are, the notches will gather moisture. Make a point of painting them with preservative as soon as you've cut them.

16

Install the notched beams temporarily using deck screws. This will allow a dry run to confirm that all of the components fit together. If you have made a mistake, you can easily make adjustments.

17

Add decorative detail to the support beams by cutting a 20-degree angle. If you have decided on an alternative style, mark and cut the ends of your two support beams now.

18

Mark the overhang of support beams by laying each beam on the ground next to the posts to measure and mark the distance you want the ends of the beams to extend beyond the posts.

Building an Arbor (cont'd)

19

Slip each support beam between the posts, and push it until it is firmly against the main beam. Line up the overhang mark, and angle a screw to temporarily fasten the beam to the post.

20

Mark for notching the crossbeams. Do not rely on measuring to mark the notches for the crossbeams—the beams may have a bit of bow in them, and your post layout may not be perfectly square. Set each one in place, and mark the notches at each end. This design calls for a 2x8 in the center of the arbor and one in from each side. The balance of the beams are 2x6s. (See the illustration on page 50.)

21

Place the crossbeams, easing them into the notches. Try to enlist a helper so that you can set both ends at the same time. With two notches joined together, fasteners aren't needed; if a notch is too wide, however, you may want to angle in a 3-in. (76mm) deck screw.

22

Install all crossbeams, working your way across the structure. In some cases, you may have to lightly persuade the beam into place. Tap on a scrap of wood to avoid damaging the wood.

23

Bore large holes for countersinking the 6 x ⅜-in. (152 x 10mm) lag screws into the support beam and crossbeam at each corner. You can surface-mount them, but countersinking and adding a plug is cleaner. Bore the larger hole first, about ¾ to 1 in. (1.9 to 2.5cm) deep.

24

Drill smaller holes to engage the lag-screw threads. The bit for the larger hole will leave a pilot hole for centering the smaller bit when drilling the holes for the lag screws. Choose a 6-in. (152mm)-long drill bit about the size of the shaft—not the threads—of the screw.

25

Fasten the lag screws without marring the beam by using a socket wrench. A touch of lubricant on the screw makes cranking it in easier. Add a washer to the screw to keep it from digging into the wood. To speed the process, buy a socket bit (inset).

Building an Arbor (cont'd)

26

Cut plugs using a drill press. (It is possible but difficult to use a drill-driver to cut a plug because the bit wants to dance off the lumber.) Once you have drilled the plugs, crack them out by levering with a screwdriver or chisel.

27

Add the plug by applying a bit of exterior glue before tapping it into the hole. Once the glue is set, use a wood chisel to trim the plug. Take off small chips until only ⅛ in. (3.2mm) or less of the plug protrudes.

28

Trim the plug flush by holding the chisel flat against the post. Try not to gouge the surface of the post. Coat the plug with sealer.

SketchUp as a Design Tool

Trying to learn a CAD (computer-aided-design) program usually ends in frustration. Many programs are overly complicated, demanding a lengthy learning curve. Don't give up. Google Sketchup™ is a fairly affordable alternative that is more than up to the job of making scale plans. With only a few hours of practice guided by YouTube tutorials, you will be able to execute a scale 3D rendering of your project. Downloadable "models" of building materials and hardware give you a leg up on developing a realistic depiction of your project.

Adding Benches

1. Cut the framing members. The frame will be 2 ft. (61cm) wide and long enough to fit snugly between the posts. Mitering the corners of the 2x6s is not essential but gives the bench a finished look. You can cut miters using a circular saw and speed square, but using a power miter saw (shown) is easier.

2. Assemble the bench frame. To eliminate visible fasteners, fasten corner brackets and small joist hangers to the end pieces. When assembling the frame, don't be discouraged if the miters aren't perfect—2-by lumber is often slightly cupped, causing the parts of the miter to show a slight gap.

3. Attach the frame. Mark a pair of posts 18 in. (45.7cm) aboveground. Measure down 5½ in. (14cm), and clamp a scrap of lumber on which the frame can rest. Level lengthwise; center the frame; and install a 3-in. (76mm) screw. Level crosswise, and install two 3-in. (76mm) screws per post. Repeat for the opposite end.

4. Cut pieces for the bench top. Confirm the width of frame, and add ⅛ in. (3.2mm) as a margin of error. Make a guide as shown to mark the bench pieces from 1x6s. Using a circular saw or power miter saw, cut enough pieces to cover both benches.

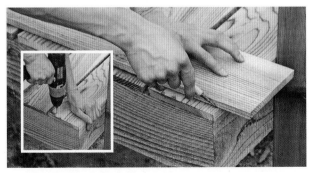

5. Attach the top pieces. Lay out the pieces, and determine how much of a gap you want. Line up the ends, and use a nail or nail set (shown) to scribe a light line as a guide for nailing. Drill pilot holes for nailing (inset).

6. Fasten the top pieces in place using galvanized 4d box or finishing nails. Use a spacer (inset), to keep the gaps between the boards consistent. Or, use 1½-in. (38mm) exterior screws, first drilling ⅛ in. (3mm) pilot holes to avoid splits.

Adding a Trellised Arbor

This arbor is designed to be strong enough to support a heavy vine and all of the fruit that comes with it while doubling as a trellis for climbing plants. But it can also function as a covered walkway, similar to a colonnade that connects a lawn to the garden or makes the transition to a shed or barn.

The arbor shown is 12 ft. (3.7m) long but can flex to suit your requirements. Should you want it longer, bear in mind that lumber longer than 16 ft. (4.9m) can be difficult to find and expensive to buy.

As with most outdoor projects, the most challenging stage is setting the posts. (See pages 80-85 for more on post setting.) Digging postholes with a clamshell posthole digger is bearable if your soil is sandy. But if you have rocky or clay-filled soil, hand digging can be a terrible job. One good option is to rent a two-person, gas-powered auger.

When built with treated lumber, the arbor will last many years without a finish. Stain, however, works better and is easier to apply than paint on treated wood. If you choose to use oak, cedar, or redwood, you will need to use stainless-steel fasteners to avoid the black stains around every bolt and screw.

Before beginning this project, contact your local building department to determine how deep you need to dig the holes to prevent frost damage.

Tools	Materials
Measuring tape	8 12-ft. (3.7m)
Line level	pressure-treated 4x4s
Mason's string	2 12-ft. (3.7m) pressure-
Plumb bob	treated 2x6s
Stakes	75 8-ft. (2.4m) pressure-
Shovel and hoe	treated 1x2s
Clamshell or other	16 galvanized carriage
posthole digger	bolts, nuts, and
Level	washers, ¼ x 5 in.
Wheelbarrow	(6 x 127mm)
Mason's trowel	1⅝-in. (41mm) deck
Power miter saw	screws
Cordless drill-driver	1-in. (25mm) deck screws
with bits	1¼-in. (32mm) galvanized
Framing square	finishing nails
Hammer	Concrete mix
Bar and spring clamps	

Trellised Arbor, Exploded View

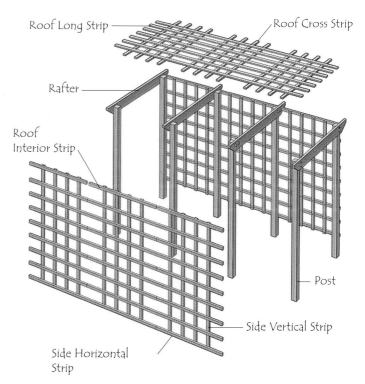

Roof Long Strip — Roof Cross Strip

Rafter

Roof Interior Strip

Post

Side Vertical Strip

Side Horizontal Strip

Combining an arbor with trellises, this handsome structure is a great space saver. It can double as a covered walkway that connects lawn to garden.

Adding a Trellised Arbor

Front and Side Views

1 **Begin the posthole layout** by installing batter boards, following the dimensions shown in the box on page 60. Drive two stakes into the ground; then add a crosspiece between them. Attach one end to the first stake; then level the other end, and screw it to the other stake.

2 **Install the opposite batter board** by attaching a string with a line level to the crosspiece on the first batter board. Hold this string on top of the second crosspiece. Move the crosspiece up or down until the string is level. Attach it to the stakes.

Adding a Trellised Arbor (cont'd)

Laying Out Postholes

Use this diagram to locate the postholes of an arbor 12 ft. (3.7m) long. To make sure the holes line up on the sides and are square at the corners, set up batter boards and string as shown. The red numbers indicate the pairs of batter boards that support a single string: 1 and 2, 3 and 4, 5 and 6, and 7 and 8.

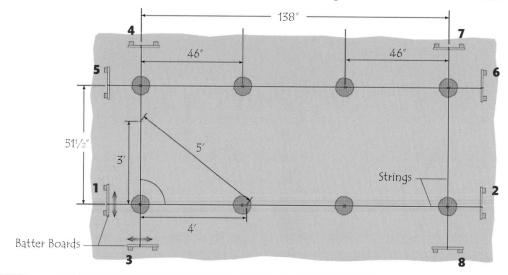

3

Establish that the layout is square by using the 3-4-5 method. Mark the 3-ft. (91.4cm) point on one string and the 4-ft. (1.2m) point on the other. Adjust the straight, taut strings until the distance between them is exactly 5 ft. (1.5m).

4

Establish the center of the arbor corner posts by lowering a plumb bob to the ground below the intersection of the end and side strings. Drive stakes at these points. Dig the holes.

5 Lower the posts into the holes, and attach braces to adjacent sides. Drive stakes into the ground about 4 ft. (1.2m) from each post. Plumb the post using a level, and attach the braces to the stakes.

6 Install all of the posts, plumbing and bracing each. As you work, make sure that the distance between the posts is 42½ in. (13m).

7 Mix some concrete in a wheelbarrow, and pour it around each post. Consolidate the concrete with a piece of 1x2 or a length of rebar. Using a mason's trowel, taper the top of the concrete so that water will run off.

8 Mark the finished post height on one of the corner posts. Stretch a string with a line level to mark the other posts; then use a square to draw a cutline at the level line. Cut off the waste using a circular saw.

Adding a Trellised Arbor (cont'd)

Setting Posts

Once you have excavated the postholes and lowered the posts into the holes, reattach the layout strings to the batter boards to establish exact locations. Move the strings 1¾ in. (4.5cm) out from their original location at the centerline of the posts so that the reference line is on the outside surface of the posts.

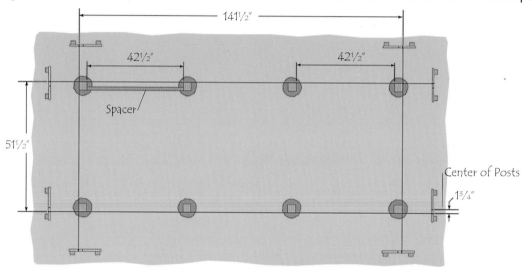

Rafter Layout

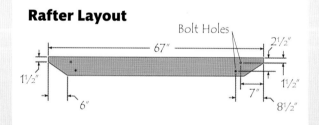

9

Lay out the size and shape of the arbor rafters using a framing square and measuring tape. Place the rafters on sawhorses or other work surface. Make the cuts using a circular saw.

10

Clamp each rafter to the sides of the posts. Lay out and drill the carriage-bolt holes through both, using a spade bit. Slide the galvanized bolts into these holes, and tighten them in place.

Top View

This view shows how the lattice strips fit on top of the arbor. Begin by installing the interior roof strips between the rafters. Then nail the short roof strips to the interior strips. Join the long roof strips to the short strips underneath using exterior screws.

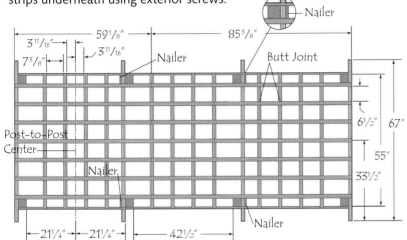

Vertical Lattice

Once the lattice on the top of the arbor is completed, install the side strips. Attach the horizontals to the outsides of the posts and the verticals to the insides of the horizontals. Use exterior screws, making sure that all strips are spaced evenly.

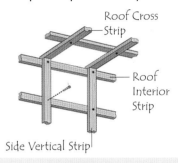

Install the nailer blocks on the sides of the rafters for fastening the interior roof strips. Attach the interior roof strips between the rafters using exterior screws. Be sure to drill pilot holes to avoid splitting the strips.

Cut the cross strips for the arbor roof lattice. Install them over the interior roof strips. Space these strips evenly using a guide made of scrap wood. Fasten the strips together by drilling pilot holes and nailing with 1¼-in. (32mm) galvanized finishing nails.

Adding a Trellised Arbor (cont'd)

13

Lay the long roof strips over the roof cross strips, and attach them using exterior screws. Use scrap pieces as spacers between the long strips. As always, drill pilot holes to prevent splitting.

14

Install the horizontal strips on the sides of the arbor posts using exterior screws. To maintain even spacing, clamp uniform scrap pieces to the posts before sliding the next strip in place.

15

Clamp each vertical strip in place so that it aligns with the roof cross strip above. Attach the vertical strips to the inside surface of the horizontal strips. Plumb the strips as shown.

16

Attach the vertical strips to the horizontal ones underneath using exterior screws. Drill pilot holes, and check for uniform spacing from side to side and top to bottom.

Building a Cucumber Trellis

Cucumbers love to climb and are equipped with little curlicue graspers for that very purpose. Give them a trellis and you can pack many more plants in your garden. Because they grow up and not out, you need only leave about 12 in. (30.5cm) between plants. You'll also harvest pristine cucumbers that have never touched the ground. All it takes is hog wire fence, 2x2 posts, and some cord. Hog wire fence panels are 50 in. (127cm) tall and 8 to 16 ft. (2.4–4.9m) long. Made of 4g galvanized wire, they'll last a lifetime. Cut them into manageable sections with a hack saw.

It is the work of minutes to pound in a few stakes each year and tie the hog fence panels in place.

Cucumbers love to grow upward, making it necessary for a climbing option like a trellis.

The tendrils of cucumbers will quickly find their way to the trellis and keep on climbing. This intensive method even allows you to squeeze in herbs like basil.

Installing a Tool-Storage Rack

Ah, garden tools. You can hang them on the wall neat as a pin, but somehow every time you pull out a rake, a shovel and hoe come tumbling down with it. Try as you might to keep things orderly, everything seems to get tangled up. What to do?

Here is a solution that takes the trouble out of tool storage. It takes a contrarian approach, organizing your tools in a rack instead of hanging them up. Inserting the tools handle first lets you pull them out with minimal fuss.

The tool-storage rack is made of two wooden grids: one on the floor and one fastened to the wall. This design can be easily custom-fitted to suit your shed or garage. The following steps show 1x3s and 1x2s, but using 1x4s or 2x2s works equally as well. Here's how to get your tools organized.

Tools	Materials
Measuring tape	1x3s
Circular saw, power	1x2s
miter saw, saber saw,	4d galvanized box nails
or handsaw and miter	2-in. (51mm) exterior
box	screws
Speed square or	Exterior glue
framing square	
Cordless drill-driver	
and bits	
Hammer	
Level	

Reuse Opportunity

Because it is out of sight and will get pretty banged up with use, this is an ideal project for using recycled lumber. The grid is only a divider with little structural function, so you can even use bits of trim. You can also rip larger dimensional lumber down to size.

Reaching for a garden tool is a pleasure with this simple rack. You set most of the tools in place handle down, but axes, mauls, and long shovels like a trenching spade can go in handle up. The design can morph to suit your location.

Installing a Tool-Storage Rack

1

Choose a location, and take measurements for the width and length of the rack. In a small garden shed, the rack can fill one wall. In a garage or larger shed, you may want it in a corner. Cut a 1x3 to length, and test-fit it (inset). Trim as needed.

2

Assemble the frame using exterior glue and 4d galvanized nails. Drill pilot holes to avoid splits. Two nails per corner does the job.

3

Test-fit the rack again once the frame is complete— it is no fun to have to dismantle the completed grid because you made it slightly too large. If the frame needs shortening, you can easily remove the nails and trim the pieces.

4

Add 1x3 dividers to the frame about every 6 in. (15.2cm). Three spaces deep is about right to allow for easy reach. Drill pilot holes, glue, and end-nail as shown.

Installing a Tool-Storage Rack (cont'd)

5

Add 1x2 dividers roughly every 8 in. (20.3cm). Drill pilot holes, and add a touch of glue at each intersection before fastening with 4d galvanized nails. Complete both frames.

6

Install the bottom frame using a 2-in. (51mm) screw. One screw is enough to hold it in place and makes removal easy for cleaning

7

Install the top frame about 36 in. (91.4cm) above the floor using 2-in. (51mm) screws driven into wall studs. Begin by leveling and fastening one end; then level across. Fasten every other stud along the length of the rack; if it doesn't run wall to wall, add a leg to one side.

8

Experiment to see which tools should go in handle down; some are easier to remove handle-up. Least-used tools, like a posthole digger, can go to the rear. Clustering the tools according to type makes it easy to find the one you want.

Making a Grow-Light Stand

A grow-light stand is indispensible if you want to get a jump on the growing season. It lets you start plants long before they are available at your local home center or nursery and allows you to start late-season plants when sets have disappeared from the market. In addition, a grow-light stand gives you the freedom to start whatever seeds you want—a boon to those who like to try heritage seeds—in an organic medium of your own choice. By blending the medium yourself and cutting seed blocks, you'll be amazed by the quick-growing, hardy plants that result. (See pages 26.)

This grow-light stand is large enough to start hundreds of plants in just 8 sq. ft. (2.4 sq. m) of floor space. It assembles quickly using inexpensive materials. Before building the stand, buy fluorescent fixtures (inexpensive utility types) and growing trays. These are the two items around which your stand needs to be built. The project allows for four 48-in. (121.9cm) fixtures and eight 10 x 20-in. (25.4 x 50.8cm) trays.

As with many projects, you'll find this one goes easier if you have a large work surface positioned at a comfortable height. A 4 x 8-ft. (1.2 x 2.4m) piece of at least ½-in.(12mm) plywood or OSB (oriented-strand board) on two 2x4s placed on sawhorses does the job.

Choose a location for the stand where dirt and water won't damage the floor. A basement or tiled room is ideal; otherwise, be sure to protect the floor from inevitable spills.

Tools	Materials
Power miter saw, circular saw, saber saw, or handsaw and miter box Framing square or drywall square Speed square Hammer Drill-driver and bits Cutting pliers	1 4 x 8-ft. (1.2 x 2.4m) sheet of ½-in. CDX or other exterior-grade plywood 4 12-ft. (3.7m) tight-knot fir 1x3s 2 12-ft. (3.7m) 2x2s 4d galvanized box nails 8 2½-in. (6.4cm) deck screws 24 2-in. (6.4cm) deck screws 4 fluorescent or LED grow light fixtures 8 fluorescent grow-light bulbs Holiday lighting timer (check wattage capacity)

A grow-light stand will give you a jump on the growing season and, once you are set up, save you serious money on plant starts. It can also be an attractive, hopeful presence when winter winds are roaring and planting time seems a long way off.

Grow-Light Stand, Elevation and Plan View

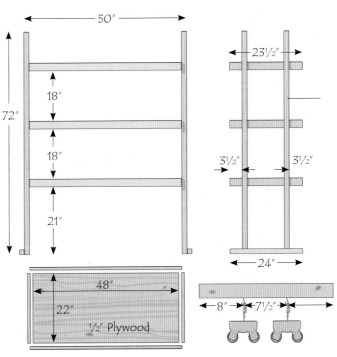

Making a Grow-Light Stand

Clamp a guide for cutting the plywood. In this case, the full 48-in. (121.9cm) width of the plywood perfectly matches the length of the grow light. Cutting four 22-in. (55.9cm)-wide pieces lets you efficiently use one 12-ft. (3.7m) 1x3 per shelf.

Mark the end pieces for each shelf. Rather than measuring, use the shelf itself as a guide to marking.

Cut the end pieces. A power miter saw (shown) is the quickest and neatest way to make these cuts, but a circular saw, saber saw, or even a handsaw and miter box will do the job.

Attach the end pieces to the plywood. Drill pilot holes to avoid splits (inset). Apply waterproof glue, and fasten using 4d galvanized box nails. The plywood may be warped a bit, so hold it down while fastening.

5

Cut and apply the long sides of each shelf using waterproof glue and 4d nails. The longer sides overlap the end pieces for a clean appearance. Secure the corners using glue and two nails as shown.

6

Cut four 2x2 uprights 72 in. (182.9cm) long and 2x2 legs 24 in. (61cm) long. Drill pilot holes, and fasten the legs using two 2½-in. (64mm) deck screws. Position each upright 4 in. (10.2cm) in from the end of the leg. Use a speed square to help position the leg.

7

Attach the shelves to the uprights. Mark the uprights so that they are located as shown on the illustration on page 71— the first 21 in. (53.3cm) off the floor, the next two separated by 18 in. (45.7cm). Clamp a scrap to support each shelf as you attach it to the uprights with two 2½-in. (64mm) deck screws.

8

Install the fluorescent or LED grow lights by hanging them from the hooks and chains provided with the fixture. Mark and predrill the underside of each shelf so that the lights are evenly spaced. (See the diagram on page 71.) Hang the fixtures, orienting the cords to the side where you want the timer.

Making a Grow-Light Stand (cont'd)

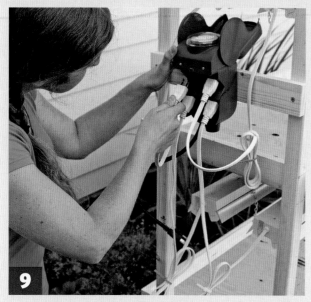

9

Install the timer onto a 1x3 attached midway between the shelves. Add cable ties to tame the fixture cords, and plug them into the timer. This timer can handle more than 1,800 watts—more than adequate for the four 80-watt fixtures.

Can Your Circuit Handle It?

Most fluorescent grow lights require 40 watts–80 watts per two-tube fixture. This stand uses four fixtures, requiring a total of 320 watts. A typical household circuit is 15 amps with a total safe capacity of about 1,440 watts. Make sure other electric fixtures and appliances on the circuit don't add up to an overload.

To check, note the wattage rating on lightbulbs and the placards on appliances. Add up all that are on the same circuit as your grow lights. Divide the total by the voltage of your circuit—120 volts. The result is the total wattage on the circuit if everything is on at once. This grow stand combined with two 100-watt light fixtures and a television, for example, might total 820 watts. Dividing that by 120 (the voltage) yields 6.83, the total safe amperage. It is well under 15 amps, so you can be confident that adding the stand to the circuit is adequate. (Aim for amperage less than 12 amps to be safe.)

10

Install the fluorescent bulbs into each fixture by inserting the pins in the socket slot and rotating the bulb.

11

Adjust the fluorescent fixture height to suit the stage of growth. As seeds emerge, lower the fixtures. Raise them as the seedlings grow. Never let the bulbs contact the plant. Set the timer (inset) to your desired lighting schedule. Pushing down small tabs programs this kind of timer.

Making Soil Blocks

The cost of started plants for a garden of any size can really add up. At $4 to $8 apiece, it is easy for the total to run to a couple of hundred dollars before you know it. And if you are buying from home centers, you don't know what went into the soil medium—of special concern if you are aiming for organic produce.

Making your own organic soil blocks, seeding them, and germinating them under grow lights not only saves money but produces amazingly healthy plants. And once you are set up, it takes hardly any more time to start your own plants than it would to drive to the nursery or home center to pick them up ready-made. (See pages 71–74 for how to make a grow-light stand and potting table.)

The right medium is at the heart of producing a healthy plant. Ray Rasmussen, an experienced four-season gardener in Washington State, begins with a blend of screened compost purchased from a local organic dairy farm, mixed with vermiculite and finely-cut peat moss. To that he adds lime and smaller amounts of amendments—kelp meal, bone meal, colloidal phosphate, and greensand.

Once you have mixed these ingredients with water, you can form them into blocks using a tool called a soil blocker and set the blocks in propagation trays. The trays have perforated bottoms so that, when set in larger watering trays, they wick up water. After seeding, put the blocks under grow lights until the plants are big enough for transplanting.

The result is a bunch of quick-growing, startlingly healthy plants that go on to flourish in the greenhouse or garden.

Seeded in soil blocks of hand-mixed medium and just the right amount of amendments, plant sets flourish under the grow lights, above right. With commercial sets costing $4 to $8 a piece, you should quickly recoup the modest cost of the necessary equipment.

A soil blocker, costing $30–$40, is essential for making your own sets, right. For another $8 or so you can buy a variety of "dibble" inserts for making the indentations into which you place seeds. If the mix is right, the blocker forms neat chunks of soil medium ideal for starting plants.

Tools

Mixing bin	Soil blocker
Scoop	Soaking tray for soil
10-quart bucket	blocker
8-oz., 4-oz. measuring	Propagation trays
cups	Leakproof trays
Screening frame	Grow lights and shelves

Making Soil Blocks

1

Sift equal portions of compost and peat moss through a screen to remove any twigs and debris. Add vermiculite and lime.

The Right Stuff

Good ingredients make all the difference in giving seeds a healthy start. You can purchase them in bulk online or from a good garden-supply store. Each has a special function. Peat is the binding agent; vermiculite, sand, or perlite keeps the mix aerated; compost gives the mix body; lime balances the PH; and the fertilizer feeds the seedlings. Here is an all-purpose blend for a large batch:

- 3 10-quart buckets of soilless potting mix (brown peat with perlite)
- 3 10-quart buckets of compost
- 1 10-quart bucket of vermiculite (or fine perlite or sand)
- ½ cup of lime
- 3 cups of fertilizer (equal parts kelp, bone meal, colloidal phosphate, and greensand)

2

Add the amendments, mixing roughly equal parts of bone meal, greensand, kelp, and rock (colloidal) phosphate. Mix them into the medium thoroughly. This is the organic fertilizer that gives the sets an early boost. For a larger batch, these ingredients should total about 3 cups. (See "The Right Stuff," above.)

3

Mix in enough water so that a handful, when squeezed, will drip slightly. The result should be firm enough to form into a loaf shape.

4

Cut the blocks. Have a pan of water handy for soaking the block cutter in advance and cleaning it between uses. Form the medium into a loaf as high as the blocker is deep. Set the cutter on the loaf, and push down firmly.

5

Release the blocks by pushing down on the handle as you hold the block cutter over the propagation tray. As the cutter forces out the blocks, it forms indentations called dibbles.

6

Place a seed in each dibble. For some crops, like lettuce, it pays to add more than one seed to ensure that you get germination.

7

Use a spray bottle to lightly spray the seeds with water. Because the soil medium is already damp, only a quick spritz or two are needed.

8

Cover the seeds with a mix of one part vermiculite to one part screened potting soil. Fill each dibble; then lightly spray the blocks again.

9

Mark the trays as soon as you plant the seeds. Have plenty of marker slips handy—often you'll have more than one kind of seed per tray.

CHAPTER 2
Fences and Pens

BUILDING A GOOD FENCE is the neighborly thing to do—and the kind thing for your livestock. Beasts that go wandering too often wind up in trouble. But building a good fence can be a royal pain. First, you must dig postholes, a knuckle-busting, backbreaking chore. Next, you need to set the posts solid and straight. Lastly, you have to wrangle fencing; livestock fencing, for one, is sharp, contrary stuff that seems to have a mind of its own.

This chapter will show you how to make the best of a tough job, even offering some easier alternatives, like the solar-powered fence on pages 106-110.

Locate All Utility Lines

Before you start digging postholes, determine the locations and depth of underground utility lines—water, sewer, gas, electrical, cable TV, and phone lines. Locating the lines that service your own house may be only half the battle. In many areas, utility companies have the right-of-way along front or back property lines for underground power cables, water mains, cable TV service, or fiber-optic transmission lines. If you accidentally cut one while digging, you could be liable for thousands of dollars' worth of damage. If you hit a power line, the consequences could be fatal.

Utility companies often will locate underground utilities free of charge. Underground lines servicing your house may be indicated on your property's original deed map or site plan. If your home was built recently, the local building department may have a record of the lines; if your home is older, added lines may not have been recorded.

If you are uncertain where such lines exist within your property, hire a private utility locating firm. Search on the Internet for "locating underground utility lines." These services usually charge by the hour. A good firm can trace and mark all underground utilities in an average-size residential lot within one or two hours.

Codes and Ordinances

Before you get too far along with planning your fence, contact the local building department to see which codes, zoning laws, and city ordinances might affect its size, design, and location. Most urban and suburban communities have fence-height laws—typically a maximum of 72 in. (182.9cm) for boundary fences in back and side yards and 36 or 48 in. (91.4 or 121.9cm) for fences bordering the street or sidewalk. In some communities, you may be able to exceed the maximum fence height if the top portion is made of wire, lattice, or some other open design.

In addition to height restrictions, codes may stipulate setbacks and easements, which require that structures be built a certain distance from the street, sidewalk, or property line. This is especially true if you are erecting the fence on a corner lot where it could create a blind corner at a street intersection or sharp bend in the road. Usually a front-yard fence more than 36 to 48 in. (91.4 or 121.9cm) high must be set back a certain distance from the sidewalk; fences more than 72 in. (182.9cm) high must be set back from side and rear property lines. Check your local codes.

Don't assume that other fences in the neighborhood meet local codes and ordinances. If your plans conflict with local zoning ordinances, you can apply for a variance: a permit or waiver to build a structure that does not adhere strictly to local property-use laws. When you apply for the variance, there's usually a fee and often a public hearing where neighbors and others involved can express their opinions. When you present your plans to the zoning commission for review, you must prove to them that you have a valid reason for requesting the variance. Even if you go through the entire process, there's no guarantee that you will be granted a variance. It's a lot easier to keep your plan with the limits of local zoning laws.

80 Post Foundations

81 Installing Posts

84 Notching Posts

86 Installing a Picket Fence

88 Choosing a Horizontal-Board Fence

89 Installing a Horizontal-Board Fence

90 Installing a Vertical-Board Fence

91 Installing a Wood-and-Wire Fence

93 Choosing Gate Latches

94 Making a Picket Gate

96 Stretching a Fence

100 Adding a Gate

106 Installing a Solar-Powered Electric Fence

111 Making a PVC Hen Pen

119 Making a PVC Hurdle

Post Foundations

Posts are typically set in tamped earth, tamped gravel, or concrete and gravel. Generally, you can use earth-and-gravel fill if the soil is not too loose, sandy, or subject to shifting or frost heaves and if the fence posts don't have to support much weight. Post-and-board fences, lattice, spaced pickets, or fences less than 60 in. (15.2cm) tall are all light enough for earth-and-gravel fill. In extremely loose or sandy soils you can attach 1x4 pressure-treated cleats to the bottoms of the posts to provide lateral stability. For long-term stability and the strength to handle just about any kind of livestock, however, concrete is the way to go.

Experienced fencing contractors pour concrete around the post, then plumb and brace it. But for most of us, it pays to plumb and brace each post securely before pouring concrete. This approach avoids disrupting the mix (or dislodging dirt) as you install the braces.

If precise post spacing is required (such as when dadoing or mortising rails into posts or attaching prefabricated fence panels or sections), you'll need to set the posts successively, fitting in rails or sections as you place each post. Use fast-setting concrete mixes for this kind of construction.

If you don't use concrete to secure the posts, it is important to compact the backfill you use, even if the substrate is gravel. The best approach is to add layers 6 to 12 in. (15.2 to 30.5cm) thick and then use a 2x4 to tamp the fill around the post.

Backfill Options

You'd normally set fence posts using a variation of one of the four basic methods shown here. Posts deeply embedded in solid, compacted soil may not need a foundation stone if they carry a lightweight fence. Corner posts and posts that support heavy gates, trellises, or other additional structures need the most secure installation.

Earth

Gravel

Gravel and Cleats

Concrete and Gravel

Installing Posts

Tools	Materials
Hammer	Wooden stakes
Posthole digger	String
Measuring tape	Spray paint
Mixing tub, hoe	Gravel
Shovel	Wooden posts
Level	Lumber for braces
Drill-driver	Concrete
Trowel	

The hardest part of installing a wooden post is digging the hole. If you have soft, sandy soil, then a posthole digger will handle the job. For many posts, renting a power auger will cut the job down to size. But rocky soil is another matter. Using a posthole digger and a 6-foot (1.8m) steel bar to loosen rocks will dig most rocky holes. Better yet, hire someone with a power auger on the back of a tractor.

1

Install layout strings that match the final location of the fence sections. If you want a square corner, make sure that the strings intersect at a 90-degree angle. Drive a stake at the intersection point to indicate where the corner post should be.

2

Mark the location of each post with a spray-paint or chalk "X," using the string as a guide. When done, remove the string and stakes.

3

Use a clamshell digger in soft, sandy soil. In rocky or clay-filled soil it makes more sense to rent a power auger. If you have a lot of fencing to install, consider hiring someone with a tractor-mounted auger.

Continued

Installing Posts (cont'd)

4

Dig to the frost line plus 6 in. (15.2cm). Check with your local building department to determine the depth. Some regions don't get cold enough to require this; in cold climates, however, setting the post too shallow will cause it to pop up and distort or break the finished fence.

5

Pour about 6 in. (15.2cm) of gravel into the bottom of each posthole to create a stable base for the post and to drain water away from the post end.

6

Replace the layout string so you know exactly where to place the post. Cut the post about 12 in. (30.5cm) longer than needed so that it can be trimmed exactly after the fence is complete. Brace the post as shown, checking for plumb on two adjacent sides.

7

Fill around each post with concrete. Use premixed dry concrete sold in bags at home centers. Mix it with water in a tub or a wheelbarrow using a garden hoe. For 10 or more posts, consider renting a concrete mixer.

8

Pour the mixed concrete into the hole around the post using a shovel. Work carefully to avoid dislodging any loose soil from the sides of the hole. Soil weakens the mix.

9

Use a long stick or the shovel handle to remove any air bubbles from the mix. An up-and-down plunging motion punctures these air pockets.

The Shock-Bar Option

When a hand-held power auger hits a root or large rock, you feel it. A more comfortable option is this hydraulic-powered auger. The hinged bar that holds the auger takes the shock instead of your arms—a big benefit when you are boring a lot of postholes. You can rent this and other powered augers, and they are inevitably worth the expense.

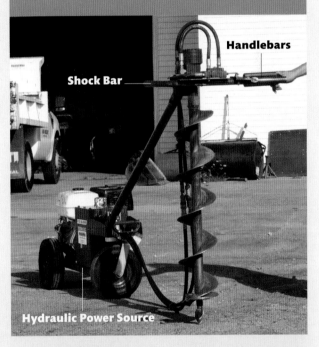

Handlebars

Shock Bar

Hydraulic Power Source

10

Make a beveled cap in the top of the concrete using a small trowel. Slant it enough so that water drains away from the post. When the concrete dries, fill the gap between it and the post with silicone caulk.

Tools		Materials	Notching Posts
Measuring tape	Circular saw	Wooden posts	
Safety glasses	Hammer	Wooden rails	
Combination square	Chisel		
Sawhorses	Mason's string		
	Line level		

Most fences are built with rails nailed to the surfaces of supporting posts. But on some fences, particularly designs with only a few widely spaced rails, you may want to dress up the installation by recessing the horizontal boards into the posts. The best tool for this job is the circular saw. By making repetitive saw cuts in the surface of the posts, you can remove the wood stock and be left with a notch that matches the size of the rail. The rails will then sit flush with the post surfaces.

1

You need to notch the post so that the rail will sit flush. Set the depth of the saw blade to match the thickness of the rail, and mark the post with two lines to match the width of the rail. Make square cuts down the middle of both lines.

2

Make repeated cuts ⅛ to ¼ in. (3.2–6.4mm) apart between the two saw cuts you made in Step 1. Try to leave wood sections no more than ¼ in. (6.4mm) wide.

3

Break the wood sections out of the notch by tapping them from the side with a hammer. If the sections are ⅜ in. (9.5mm) thick or more, you may find it easier to run the saw through them again and crack them out.

4

Smooth the bottom of the notch using a hammer and sharp chisel. Test-fit a scrap piece of the rail stock in the notch. If the notch is too tight, make another cut or two.

5

Lower the posts into their holes, and check for alignment with the notches in other posts using a mason's string and a line level. When you are satisfied with the alignment, plumb and brace the posts.

Bevel-Cut Posts

If you want decorative detail on the tops of the posts, you can bevel-cut the edges using a circular saw. Make a small cut along the edge, or make a deeper cut to create a wider bevel. It is not easy, but with practice you can even turn the top of the post into a pyramid shape. For a slightly beveled edge, use a block plane.

Rail-Connecting Alternatives

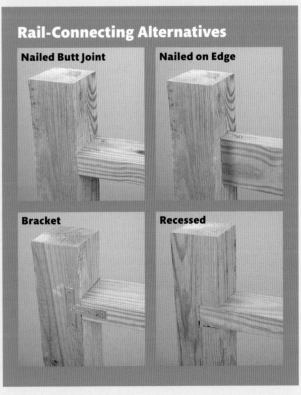

Nailed Butt Joint

Nailed on Edge

Bracket

Recessed

Installing a Picket Fence

Tools	Materials
Drill-driver, clamp	1x4s for pickets
Plumb bob, string	2x4s for rails
Saw	4x4 posts
Hammer	Scrap for spacers
Line level or 4-ft.	1½-in. (38mm)
(1.2m) level	exterior screws
Measuring tape	

While not ideal for livestock, a picket fence is a beautiful way to sequester a garden or orchard. And once you get the posts installed and trimmed to the desired height, picket fences are actually fun to install. A common rule of thumb is to use a space that's the same width as one of the pickets, but you may have to vary the gap to even things out along the run of the fence. To determine the finished picket spacing, first measure the distance between the posts. Then divide this by the width of a picket and of a single space. If you don't come up with a whole number, then slightly increase or decrease the width of the spaces. Rip a piece of 1-by as a spacer when installing the pickets. Check for plumb every few pickets.

1

Install the rails. If you have notched the posts (pages 84-85), clamp the rails in place; drill pilot holes; and drive exterior screws or nails through them into the posts. Or use one of the methods shown on page 85.

2

Install a guide string between the posts to keep the picket tops level as you attach them to the fence rails. Run the line along the top or bottom of the fence, whichever you find most convenient. Draw the string tight, and check for level using a line level. Install the pickets so that they fall just above this line.

3

Cut a scrap wood spacer once you establish the proper spacing between pickets. Use the spacer to align each successive picket. Fasten the pickets using 1½-in. (38mm) exterior screws.

Post-Top Variations

There are many ways to top posts aside from a basic square cut. Some of the best options combine decorative details with the ability to shed water, which can shorten post life by causing rot where it col-lects on the porous end grain. The easiest approach is to make an angled cut. Chamfering the top edges also helps. Adding a full cap rail, either chamfered or angled, also reinforces the fence.

Angle Cut

Chamfered Edges

Flat Cap with Chamfers

Angled Cap

Picket Styles

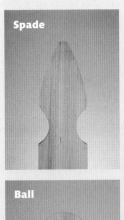
Spade

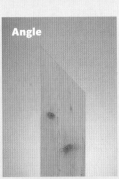
Angle

Spear

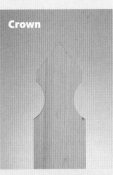
Crown

Crown and Ball

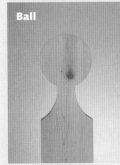
Ball

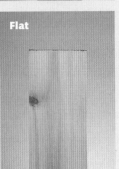
Flat

Provincial

Round

Point

Choosing a Horizontal-Board Fence

The most basic horizontal-board fence uses a framework of 4x4 posts set 8 ft. (2.4m) on center and two or more rows of 1x6 rails. You can dramatically alter the appearance and function of this basic fence by using three or four horizontal rails, among other options. If necessary, you can add a 1x6 vertical board nailed mid-span between posts to reinforce the horizontal rails and minimize bowing. Among other options, you can use a variety of board widths to make a repeating pattern or place a wide board near the ground and progressively narrower boards above.

You might also add a cap rail over the posts of a board fence (previous page). The best kind of cap has a slight slope, which requires cutting the post top at an angle. If you attach a cap rail without angle-cutting the post, place the cupped side up so that water drains off.

When you install the rails (and a cap), be sure to stagger joints on different posts on a long run. You can use rough-sawn lumber or even siding boards that match your house.

Alternative Horizontal-Board Styles

The most economical pattern, and the easiest to install, is the basic four-board fence (below). For more protection you can add a mid-board or several boards with small spaces between them. Another twist on this idea is to use different widths to create a pattern (right). Another classic board-fence pattern is the basic X-shape between posts, used in combination with top and bottom rails (bottom right).

Patterned Board

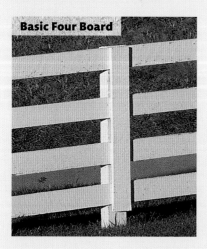

Basic Four Board

Cross Board

Installing a Horizontal-Board Fence

Tools	Materials
Mason's string	1x6 rails
Line level	4x4 posts
Measuring tape	2-in. (51mm) exterior
Safety glasses	screws
Saw	
Hammer, chisel	
Drill	
Block plane	

Cut the notches in the corner posts beforehand on a worktable or sawhorses. Install the corner posts first. Use a temporary crosspiece (inset) to set them into postholes at a consistent height. Use mason's string with a line level as a guide to marking for notches in other posts. Set the posts, and cut the notches using the technique shown on pages 84-85.

A horizontal-board fence is one of the simplest designs around, an offshoot of the standard rustic fence. The posts are widely spaced for quick construction, and there are no rails, so the fence is less expensive to build. Horizontal fences are generally built to conform to the rise and fall of the ground. Often, wire fencing is added to make the fence suitable for smaller livestock. (Also see pages 92-93.)

Join the fence boards at the center of a post notch. Avoid splitting the boards by drilling pilot holes. If a board is slightly too big for the notch (lumber width and thickness can vary slightly), give it a few swipes with a block plane until it fits.

Attach the boards using exterior screws. Two per board should be enough. If the board is twisted and won't sit flat in the notch, add a screw to force it into place.

Installing a Vertical-Board Fence

Tools

Measuring tape
Hammer
Level
Safety glasses
Saw
Cordless drill-driver
  (optional)

Materials

2x4s for rails
1x6 boards
Rail hanger hardware
8d galvanized nails
4d galvanized box
  nails
Exterior screws (opt.)

1

Install the rail hangers by establishing the height of the bottom rail, centering the hanger on the post, and then fastening it with 8d galvanized nails. Cut the rail to length, and slide it into the hanger. Level the rail across to the next post, and mark for fastening a hanger. Fasten the hanger; slip in the rail; check for level; and fasten using 4d galvanized box nails.

If you want a degree of separation from your fence but don't care for the wall-like effect of boards set side-by-side, try this approach. Alternating boards on each side of the fence creates a barrier but maintains a pleasant degree of openness. This project uses hanger hardware to install the rails, which is not only quick but allows you to install the rails dead center on the posts. This is a forgiving fence to build: if the extent of overlap varies a bit, you won't notice it.

2

Install the top rail hanger the same way you installed the bottom one. Then slide the rail into place, and check it with a level. When you are satisfied with the positioning, nail the rail to the hanger.

3

Calculate the best spacing between the boards, and cut a scrap board to match the width of a single space. Use the spacer as a layout tool to mark each board location. Fasten the boards using 4d galvanized box nails.

4

Install all of the boards on one side, and then work on the other side. Every fourth board or so, use a level to check for plumb. (If you prefer, you can use a drill-driver and exterior screws instead.)

Installing a Wood-and-Wire Fence

Tools	Materials
Shovel	¾-in. (19mm) U-staples
Hammer	2-in. (51mm) screws
Drill-driver	Wire fencing
Electrician's pliers	1-by for post covers
	Pressure-treated 1x6
	skirt board

1

Dig a trench to bury the wire to keep varmints out. To ensure protection, bury about 12 in. (30.5cm) of fencing.

2

Plan to use the full height of the wire between the ground and the top of the top rail. This eliminates cutting along the bottom and lets you line up the edge of the wire along the top rail so that you'll get no sagging between posts.

Continued

Welded- or woven-wire mesh on posts and rails makes an economical, easy-to-build fence ideal for backyard homesteads. Extend the posts, and this design serves well as a trellis for climbing vines. Installing wire fencing without rails saves some steps and material but requires fence stretching, an art onto itself. (See pages 96-99.)

Welded wire comes in a variety of gauges and mesh sizes, typically in 36-, 48-, and 72-in. (91.4, 121.9, 182.9cm) heights, in 50- or 100-ft. rolls. Choose the heaviest-gauge wire available: thin wire can easily deform and is prone to rust. Most wood-and-wire fences have a 2 x 2-in. (5.1 x 5.1cm) or 2 x 4-in. (5.1 x 10.2cm) galvanized or vinyl-coated grid.

Two rails will do on fences no more than 48 inches high, but you'll need a mid-rail for extra stability on higher fences. You can enhance the fence with two add-ons: an angled cap rail that sheds water and covers the wire tips, and a pressure-treated skirt board that keeps critters from crawling under the fence. You can backfill against this board with topsoil.

Deer Defense

To discourage deer, you can build a simple wire fence attached to a wooden post-and-rail framework that is 8 ft. (2.4m) high, the recommended height. Because deer can't jump high and long at the same time, however, a 6-ft. (1.8m) fence leaning outward at a 45-degree angle is equally effective. Electric fences help, too, especially if you train deer to avoid them by luring them with peanut butter to give them an educational shock or two. (See pages 106–110.)

Installing a Wood-and-Wire Fence (cont'd)

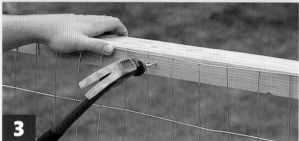

3

Attach the leading edge to the first post using ¾-in. (19mm) U-shape staples (top). Use the same staples to attach the top edge of the wire to the top rail (bottom). Align the top edge of the wire with the top of the rail.

4

Cut the wire flush to the outside edge of the last post using wire cutters or electrician's pliers. If you have to splice wire rolls, do it over a post and overlap each roll by one wire grid. Use more staples rather than fewer to make the joint strong.

5

Reinforce the joint between rolls where needed by covering the post with a ¾-in. (19mm)-thick pressure-treated board. Use exterior screws to attach the board. To maintain a uniform appearance, you may want to cover all of the posts with the same kind of board.

6

Install a skirt board along the bottom of the fence to reinforce the lower rail and the wire that extends into the trench. This provides increased protection against burrowing animals. Once you have installed the board, backfill underneath.

Choosing Gate Latches

The first rule of farming is "Latch the gate behind you." In the old days, a bit of wire served as a latch, but there are many more-substantial kinds of hardware from which to choose. They range from a basic hook-and-eye, which may be inconvenient on a gate that's used frequently, to hasps that you can secure with a padlock—overkill in most cases. The most useful are sliding bolts and traditional wrought-iron latches. You can install some kinds in combination with a spring-loaded hinge to close and latch the gate.

Another approach is to build your own custom latch, like the one shown below. (For another handmade option, see page 105.) This latch is made of 1x2 pine, including a bolt that slides through three "keepers." You fasten two of the keepers to the gate and one to the adjacent post. With a dowel or some other type of handle between the two gate-mounted keepers, you can slide the latch back and forth across the opening to the post keeper. Because the handle protrudes, the latch won't slide free.

Making a Sliding Latch

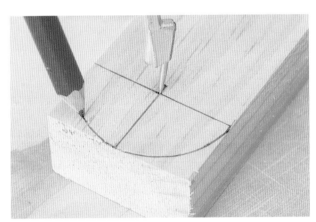

1. Scribe an arc to round the end of the sliding latch board. Sand the edges to keep it from catching as it slides.

2. Cut the arc on the end of the latch board using a saber saw.

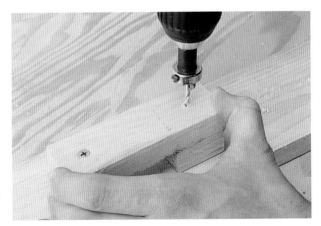

3. Make keeper assemblies by predrilling and fastening a cover piece on blocks the same thickness as the latch.

4. Level the three keepers (two on the gate and one on the post); slide in the latch board; and add a dowel handle.

Making a Picket Gate

1 **Choose a frame corner joint** from those shown on the opposite page. Then cut the sides of the frame to size using a circular saw. Align the corners; drill two pilot holes in the overlapping board; and fasten with 2½-in. (64mm) exterior screws.

Tools

Drill-driver
Safety glasses
Screwdriver bit
Measuring tape
Chisel
Circular saw
Framing square

Materials

Pickets
Metal L-brackets
2x4s
2½-in. (64mm)
  exterior screws
1½-in. (38mm)
  exterior screws

Constructing a picket-gate frame is a straightforward job if the opening between the posts is parallel and plumb. Start with a square frame, and add diagonal bracing inside. If the posts are not parallel and plumb, don't try to make the gate fit a bad opening. Fix the posts instead by either installing sag cables to keep them aligned or resetting the posts.

2 **Reinforce each corner joint** with a galvanized-steel L-bracket. First, fasten one leg of the bracket to a frame side. Then, hold a framing square against the corner to align the sides. Screw the second leg in place.

3 **Install a diagonal brace** by cutting a 2x4 slightly longer than needed and placing it on the gate frame so that it extends from the middle of one corner to the middle of the opposite one. Mark a V-shape cut on both ends. Make the cuts, and slide the 2x4 in place.

4

Lay out a half-lap joint in the middle of the diagonal braces. Cut the joints using a handsaw and chisel. Then push the braces together inside the gate frame.

5

Fasten the braces by spreading exterior-grade construction adhesive on both sides of the joint. Drill a pilot hole, and drive a 2½-in. (64mm) exterior screw into the middle of the joint.

6

Lay out the pickets (or other boards) across the gate frame to establish the best spacing. Then install the pickets using 1½-in. (38mm) exterior screws. Drill pilot holes for the screws to prevent splitting.

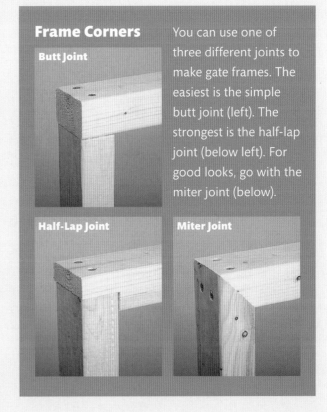

Frame Corners

You can use one of three different joints to make gate frames. The easiest is the simple butt joint (left). The strongest is the half-lap joint (below left). For good looks, go with the miter joint (below).

Butt Joint

Half-Lap Joint

Miter Joint

Stretching a Fence

Of all the backyard homesteading chores, running fence has to be the universal least favorite. From digging postholes to wrangling uncooperative fence wire, it is a succession of wearisome tasks. An electric fence is a labor saving alternative. (See pages 106–110.) But for the perimeter of your property and for a barrier that suits a variety of livestock, a wire fence is the way to go.

Next to a well-dug post set straight and true in concrete (pages 81–83), quality stretching is what a good fence is all about. If you have ever tried to stretch with brute force alone, you know the meaning of futility. Make a poor job of it, and the wire fencing sags, buckles, or sits too high above the ground. Your livestock will find a way out, endangering them and wasting all of your effort.

The following steps show a tried-and-true method used by veteran farmer Ernie Schmidt. Key to his method is a tool called the come-along. This handy device ratchets cable or chain, exerting a mighty pulling force. Own one, and you'll be surprised at the jobs it can tackle: pulling out bushes, straightening walls, rescuing a car stuck in the snow. For fence stretching, it does what the human back cannot do—pull and hold a couple of hundred feet of wire fencing nice and tight.

WARNING: a come-along can be dangerous. Make sure the posts at both ends of your fencerow are strong and stable. Secure the fencing well so that there is no chancing of its ripping loose as you stretch it. As you work the come-along, do so slowly, watching for any strains.

Tools	Materials
Come-along	Wire fencing
2 chains with hooks	Fence staples
2 large C-clamps	Baling twine
Hammer	2 4-ft. (1.2m) 2x4s
Measuring tape	Lumber for bracing
Pliers	Lumber for levers
Cutting pliers	
Locking pliers	

Gather together the tools and materials for the job to avoid annoying runs back to the store to shop for needed items, above right. A strong come-along is the essential tool: it exerts the pulling power to stretch your wire fencing drum-tight.

A wheelbarrow makes a handy tool carrier for trundling everything you need out to the fence site, right. As with any job, it pays to bring along everything you can imagine needing.

Stretching a Fence

1

Roll out the fencing along the row of posts. Continue rolling beyond the last post so that you can easily stand up the roll.

2

Staple one end of the fencing, letting about 4 in. (10.2cm) of wire hang beyond the post. Pound in a heavy-duty staple; bend the wire back over the staple; and add a medium-size staple. Work your way down the post, securing every horizontal strand in the same way.

3

Temporarily tie the wire fencing to the posts so that it is held upright. Tie the fencing to every third post or so. Baling twine—the farmer's friend—is ideal for this. Work your way down the fencerow.

4

Sandwich the fencing between 2x4s using large C-clamps. The 2x4s grip the fencing without damaging it. Fold the roll out of the way.

Continued

Stretching a Fence (cont'd)

5

Hook the chains around the clamps. Wind a second chain around the opposite post for anchoring the come-along. In this instance, a gate will eventually go between the posts, so they are in a handy position for stretching.

6

Add bracing to reinforce the posts, in this case a bridging 2x4 between the post tops. In other circumstances some 2x4s placed diagonally may be necessary. Do whatever it takes to ensure that you have fastened the come-along to a secure base.

7

Crank on the come-along until the fencing straightens up. Walk along the fencerow to check for any strains or loosening. If all looks well, continue cranking the come-along.

8

Begin stapling the fencing after you pull it as tight as it will go. You should be able to pluck the fencing and see it vibrate. Add a staple at every horizontal wire all the way up the first post. Secure the bulk of the fencing by stapling every other post.

9

Drive a lever to force the fencing against the post before stapling. The ground is never level, so you may have hold the fencing down with your boot to force it closer to the ground.

10

At the terminal post, wrap the wire around it; cut the fencing, leaving enough excess to pull individual wires all the way around the post. Bend a hook in the end of each wire, and to clip it over the stapled fencing.

11

Staple each post once the wire fencing is secure at both ends so that there is a staple at the top, the bottom, and every 12 in. (30.5cm) in between. Once you have completely stapled the wire fencing, loosen the come-along and remove any temporary bracing.

Adding a Gate

A farm gate needs to be strong enough for a child to swing on it but not so beefy that it strains the post to which it is attached. That means attaining the right balance between structural integrity and strength. This simple gate hits it just right.

Made of 5½ x ⅝-in. (14 x 1.6cm) cedar fence boards on a 2x4 frame, this gate goes together in a couple of hours. It owes its stiffness to the Z-bracing of the 2x4s and the careful application of four galvanized nails at every overlap of the frame and the boards.

To prepare for installing a gate, always make sure that the adjacent posts are set deeply and braced well. Veteran farmers assert that a gate is only as strong as the post to which it is attached. Its posthole should be a minimum of 3 ft. (91.4cm) deep—especially important for the post to which the hinges are attached. A 4x4 post is easier to attach hinges to than a round post. Run the fencing up to the latch side of the gate, but leave it off the hinge post so that you can firmly attach the hinges without having the fencing get in the way. When building a gate with a Z-brace, always run the diagonal piece from the topmost hinge down to the opposite bottom corner. That way, the gate structure hangs from the hinge.

Tools	Materials
Measuring tape	5½ x ⅝-in. (14 x 1.6cm)
Circular saw	cedar fence boards
Square	2x4s, 2x2
Hammer	5d (1¾-in. [45mm])
Clamp	galvanized box nails
Drill-driver and bits	3-in. (76mm) exterior
Saber saw (optional)	screws
	Gate latch
	Utility hinges

You can adapt almost any lumber to this design, but be cautious of making the gate too heavy. The 5½ x ⅝-in. (14 x 1.6cm) cedar fence boards shown are inexpensive and attractive—light but solid enough for a goat to rub against without compromising structural integrity. Cedar is easy to cut, but be sure to drill pilot holes when fastening within and inch or two of the end of a board.

Simple, lightweight, and attractive, this gate goes together in a couple of hours and will do the job of corralling beasts for decades. A 2x4 Z-brace bears the structural load; 5½ x ⅝-in. (14 x 1.6cm) cedar fence boards finish the job.

Adding a Gate

1

Measure the opening at the top and bottom. Chances are that the measurements will differ. Choose an overall measurement that is about 1 in. (2.5cm) less than the smaller dimension.

2

Cut two 2x4s to suit the opening. Cut the fence boards to the desired length; 4 ft. (1.2m) is a good choice for most livestock. Lay the 2x4s on a smooth surface, and place your fence boards on top of it. Use a spacer board to rough out the gap between the boards.

3

Determine the distance you want the boards to overrun the frame: in this case, they extend 6 in. (15.2cm) above and below it. Position the first and last boards. Nail a 5d galvanized box nail where the boards overlap the frame.

4

Measure the diagonals to square up the gate. Both measurements should be equal (inset). Make any needed adjustments. When you are satisfied, pound three more nails where the each board overlaps the frame.

Continued

Adding a Gate (cont'd)

5

Use a 2x4 guide to position the boards as you fasten them. A concrete block helps hold it in place.

6

Fasten the boards to the frame, using the guide board and spacer in combination. Use four 5d nails at each intersection with the frame. Check your work midway (inset) to be sure you are not compounding any warp in the boards.

7

To avoid splits, work around any knots or drill pilot holes before nailing. Try to stagger the position of the nails along the grain to eliminate cracks. Nail all of the boards in place using four nails at each intersection with the frame.

8

Position the 2x4 cross brace, and using a framing square, mark it for cutting.

Add a Decorative Touch

Use a pint can to mark a scalloped decoration. Make the cut using a saber saw. Other options include nipping off the corners, angled points, or diagonal points. To mark a round cut on the top of each board, use a large juice can.

 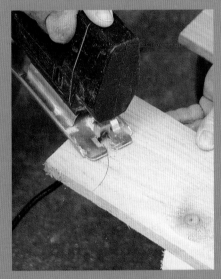 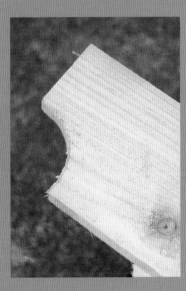

9

Make the angled cut in the 2x4. Clamp the board before working to ensure a straight cut.

10

Fasten the diagonal brace to the 2x4 frame using two 3-in. (76mm) screws. Angle the screws to firmly fasten the pieces together. That means that the angles of the screws differ from each other (inset).

Continued

Adding a Gate (cont'd)

11

Pound four nails into each overlap of the boards and the angle brace, spacing them to avoid splits. Add a single nail (inset) to the end of the brace.

12

Attach the hinges directly over the frame. Drill angled pilot holes to avoid splitting the 2x4.

13

Attach the gate hinges to the post. Use scrap lumber to position the gate. Aim for a minimum of 2 in. (5.1cm) and a maximum of 6 in. (15.2cm) off the ground. Fasten one screw per hinge, and give the gate a test swing to be sure it clears all obstructions.

14

Dig away any high spots, and test the swing again. Or you may choose to raise the gate higher on the post.

15

Install a 2x2 stop to the latch post using 3-in. (76mm) deck screws. Before fastening, position the 2x2 so that the front of the gate is even with the front of the latch post.

A Handmade Alternative to Hardware

This simple latch operates easily and, because it rotates rather than slides, won't bind up in wet or icy weather. Made of 1-by lumber, the stop board for the gate positions the rotating latch so that it catches readily.

16

Hold the latch to the gate, and position it so that the bolt easily reaches the catch. Attach the latch first, then the catch.

17

Bend the end of the hinge around the post by tapping it with a hammer. Add a screw to every hole in the hinge.

Installing a Solar-Powered Electric Fence

An electric fence is a simple, effective confinement solution for livestock. It is essentially a psychological barrier; animals quickly learn to avoid those unpleasant wires. An electric fence does not need to be built as strongly as a traditional one. That means no posthole digging, wire stretching, or corner reinforcement.

Using metal posts and a few strands of wire, you should be able to install an electric fence quickly. A solar-powered electric fence like the one shown eliminates the need for running electrical power to the fence through an underground cable or using a lead-acid battery—the major complications of electric fence installation. Instead of trenching cable, it requires only attaching a solar-panel-equipped energizer panel to a fence post.

It has its limitations, however. While ideal for penning off animals—from chickens to hogs—within your property, it should not be used as the final perimeter barrier that guards livestock from wandering onto roads or neighboring property. For the actual perimeter you should double up, combining an electric fence with traditional fencing.

Good grounding conditions are essential for an electric fence. A dry climate, frigid winters, or rocky soil reduces the effectiveness of grounding. Check the manufacturer's recommendations for adding additional grounding rods in such conditions. Solar installations require a bit more maintenance than that for standard electrical fences. For example, pasture areas can get dusty, so check the solar panel periodically, and clean it as needed. During long cloudy periods, the solar panel may not be able to recharge the battery. Check the digital voltage meter to confirm that the battery is charged. If necessary, remove the battery and follow manufacturer's instructions for recharging it using a trickle charger.

An electric fence, left, is a simple way to provide a psychological barrier that safely confines wily creatures like goats. Even a couple of strands added to an old wire fence can restore security. Some livestock owners train their animals to shun electric fences by running a charged wire across the yard and placing tempting feed on the other side of the wire. The sting of a few shocks will teach them to avoid any fence.

Orient the solar panel so that it faces due south, below left. Choose a location away from trees or buildings that could put it in shadow.

Tools	Materials	
Post driver	Steel T-posts	galvanized-steel or
Measuring tape	Plastic brackets	copper grounding
Cordless drill-driver	14-ga. galvanized-	rod with clip
with bits	steel or copper wire	Insulating tubing
Ratchet-action socket	Screw-in wire holders	Energizer panel
wrench and sockets	with insulators	Wire-splicing sleeves
Electrician's pliers	6–8-ft., (1.8–2.4m)	(optional)
Grass/weed trimmer	⅜–⅝-in. (10–16mm)	

How It Works

The solar panel charges a battery inside the energizer panel. When an animal touches the electric fence, the panel releases a very short pulse of high voltage. The shock current flows through the animal to a ground wire in the fence or through the soil itself. It eventually reaches a ground rod and is conducted back to the energizer to complete the circuit.

The essentials of a solar-electric-fence system: aluminum wire, screw-in wire holders with insulators for wooden posts, clip-on wire holders for steel posts, a solar energizer panel.

Solar-Powered Electric Fence

A multi-strand electric fence includes grounded wires to ensure a shock even in very dry conditions. Electric fences can be configured to suit your type of livestock—from chickens to hogs.

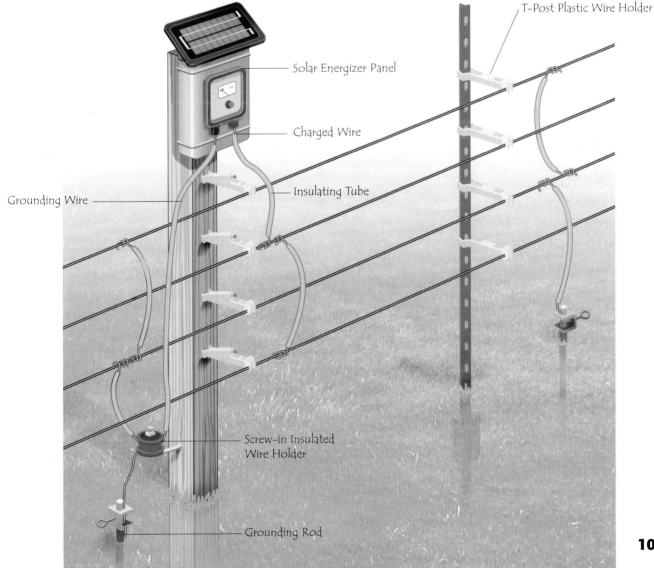

T-Post Plastic Wire Holder

Solar Energizer Panel

Charged Wire

Insulating Tube

Grounding Wire

Screw-in Insulated Wire Holder

Grounding Rod

107

Installing a Solar-Powered Electric Fence

1

Pound in the pointed end of the steel T-posts. Orient the bumps of the posts where you want the plastic wire holders to protrude. (See Step 2.) Use a post driver (shown); it is fast, easy, and safe. Drive the post at least 12 in. (30.5cm) into the soil. Post spacing can range from 10 to 50 ft. (3–15.2m).

The Importance of Being Grounded

An electric fence is only as good as its ground. That involves pounding a 6- to 8-ft. (1.8–2.4m), $^3/_8$- to $^5/_8$-in. (9–16mm) copper or galvanized-steel rod into the earth and connecting it, directly or indirectly, to the energizer panel. Check manufacturer's instructions for installing rods in excessively rocky or hard soil by pounding in several rods at an angle and connecting them together. Dry soils require more rods at more-frequent intervals. Use 14-gauge wire to connect grounds. Use a screw-type connector to attach wire to the grounding rod.

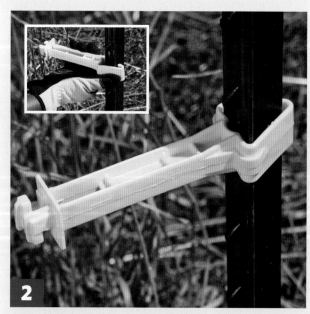

2

Attach the plastic wire holders to the T-posts by hooking them sideways on the post (inset) and snapping them into place. Position the wire at the nose height of the animal you are confining. If you plan to include several kinds of livestock, position more than one wire.

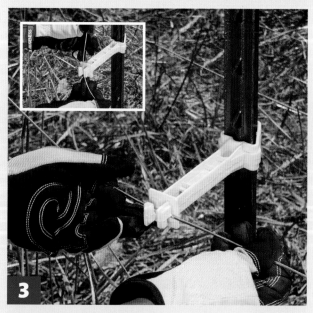

3

Attach the charged line to the holders by inserting the wire into the holder vertically (inset) and then clipping it in place.

Continued on page 110.

Attaching Wire Using Metal Holders

A **typical electric-fence installation** combines steel posts for partitions within a field and wooden posts with hardware-wire fencing—augmented by electric fencing—around the perimeter. You'll need metal screw-in wire holders with insulators wherever the tension on the wire might be too much for a plastic holder. (See Step 4, page 110). Other typical locations needing metal holders include corners (where the wire turns 90 degrees) and on the post below the energizer panel. You might also use metal screw-in holders when adding a charged wire to bolster the effectiveness of an existing fence. Plastic wire holders serve for straight runs where tension is minimal.

1. Drill a pilot hole for the holder at the needed height. Choose a drill bit of a diameter slightly less than that of the holder's threads.

2. Screw in the holder using pliers or the claws of a hammer. Slip on an insulator and nut.

3. Fasten the insulator in place using a socket wrench.

Installing a Solar-Powered Electric Fence (cont'd)

4

Use a nail-in plastic wire holder where the wire runs straight or is otherwise not under tension. Once you have installed the wires, remove weeds to keep them from discharging the system. A grass/weed trimmer makes quick work of the chore.

5

Connect the charged and grounding wires for final hookup to the energizer panel. Bend the connecting wire to cross over adjacent wires, and add insulator tubing to the connection as shown. You can twist wires together or attach them using wire-splicing sleeves as shown. Attach two screw-in wire holders to the post that holds the energizer panel for incoming grounding wires (inset).

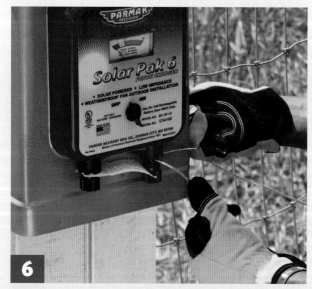

6

Install the energizer panel, fastening it with screws through the mounting holes in the back of the unit. Switch off the energizer, and attach the charged line to the red terminal on the panel. Twist the wire around the connector shaft, and tighten firmly.

7

Attach the grounding line to the black terminal of the energizer panel. Before releasing livestock into the pasture, allow five full days of sunlight for the battery to charge. Or charge it using a trickle charger.

Making a PVC Hen Pen

If the full-size chicken tractor on pages 142–150 is more equipment than you need, consider this simple PVC pen. It is easy to make and light enough to drag from your coop to a grazing area—with the chickens shuffling along inside. Features include a swinging flap for easy access and an attached tarp to shield the layers from sun or rain.

This project requires lightweight netting. Bird netting is too fragile; instead, use plastic hardware cloth. It is light but tough—just right for the job.

Working with PVC pipe is simplicity itself. You can cut the pipe using any fine-tooth saw and "weld" fittings using primer and cement. Working together, the primer and cement break down and melt the mating surfaces of the PVC; once dry, the fitting and pipe are firmly bonded. That is why PVC pipe has been so successful as a plumbing material. It is also forgiving. If you don't get the assembly quite right, there is usually enough "give" in PVC to make up the difference.

Tools	Materials
Measuring tape	8 10-ft. (3m) pieces of
Black felt-tip	½-in. (13mm) PVC pipe
marker	14 ½-in. (13mm) T-fittings
Fine-tooth saw	4 ½-in. (13mm)
Miter box	90-degree elbows
Coarse sandpaper	1 ½-in. cross-fitting
Speed square	PVC primer (clear, not
Sliding adjustable	purple, if available)
bevel	PVC cement
Drill-driver and bits	2 2-in. (5.1cm) eye bolts
Scewdriver	with nuts
(optional)	3 rolls of 3 x 15-ft.
Adjustable wrench	(91.4cm x 4.6m)
Aviation shears or	plastic hardware mesh
heavy scissors	140 6-in. (15.2cm) cable ties
Cutting pliers	1 5 x 6-ft. (1.5 x 1.8m)
Grommet kit	plastic tarp (min.)
	3 #10-32 x 2½-in. (81.3 x
	6.4cm) machine screws

This lightweight pen is a great way to give your layers a field trip. Just park it next to your coop; entice the ladies into the pen; and slide the pen, chickens and all, to a prime grazing location.

Hen Pen Exploded View

Use this diagram as a guide for building the pen. It easy to go wrong, so check twice before welding the components. Cut pieces only as you need them; it is otherwise too easy to mix them up.

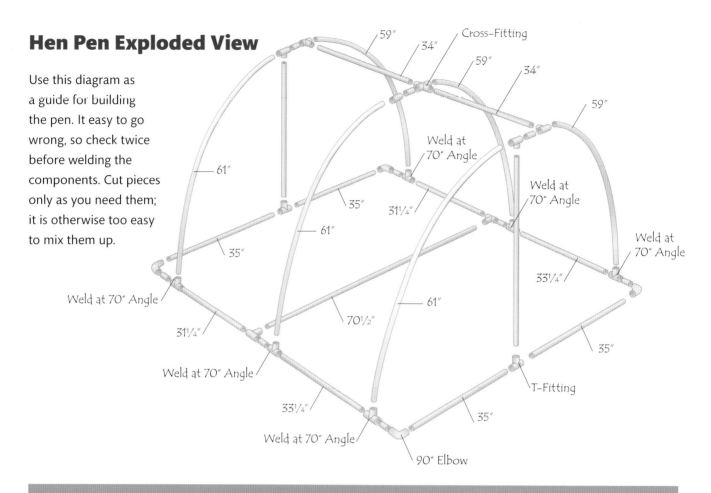

59" 34" Cross-Fitting

59" 34"

59"

61"

Weld at 70° Angle

Weld at 70° Angle

Weld at 70° Angle

35" 31¼"

61"

35"

61"

33¼"

Weld at 70° Angle

31¼"

70½"

35"

Weld at 70° Angle

33¼"

T-Fitting

35"

Weld at 70° Angle

90° Elbow

The Method

1. Wipe clean the pipe and fitting.

Correctly done, a PVC joint is as strong as the pipe itself. The primer and cement softens the pipe and fitting so completely that once the joint dries, the two are as one—the reason this technique is know as PVC welding. The process is simple and straightforward: clean; prime; cement; then push and twist.

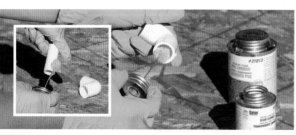

2. Use the primer applicator to coat the pipe and the inside of the fitting. Before the primer dries, apply the cement to the pipe (inset) and inside the fitting.

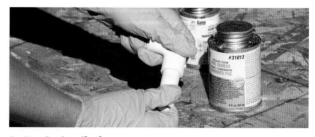

3. Push the fitting onto the pipe, giving it a one-quarter-turn. Let critical joints dry for at least a couple of hours.

Making a PVC Hen Pen

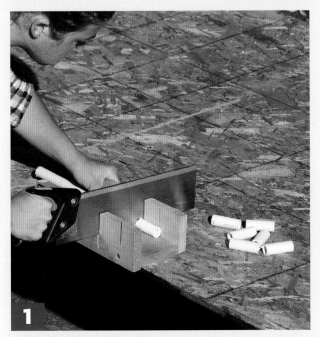

1

Cut nine 2-in. (5.1cm) pieces for joining the fittings. Measure the required length, and mark the cut line using a black marker. Using a miter box ensures a neat, straight cut.

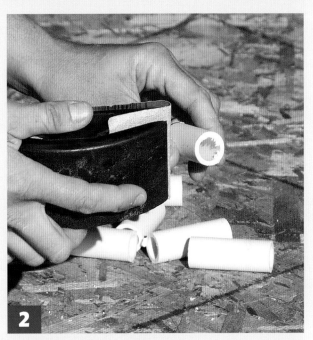

2

Deburr all of the pieces using coarse sandpaper, and wipe off all debris using a clean rag.

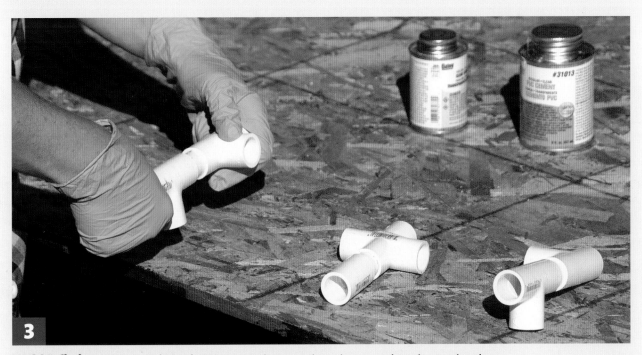

3

Weld T-fittings with 2-in. (5.1cm) pieces to make two mirror-image end sections, using the illustration on page 112 as a guide. Join a straight fitting and the cross-fitting to make the center segment. Weld them using the PVC joining process shown in the "The Method," opposite.

Continued

Making a PVC Hen Pen (cont'd)

4

Cut three 59-in. (149.9cm) pieces from the 10-ft. (3m) pipes, and weld them to each of the assemblies made in Step 3.

5

Weld the 61-in. (154.9cm) cutoffs remaining from the 10-ft. (3m) pieces of pipe to the opposite ends of the three assemblies you've completed.

6

Cut two 43½-in. (110.5cm) pieces, and weld them to the T-fittings of the two end assemblies made in Step 5.

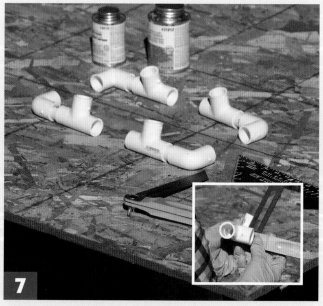

7

Make the corner assemblies for the base of the pen, combining a T-fitting, 2-in. (5.1cm) piece of pipe, and 90-degree elbow. Use the speed square to set the sliding adjustable bevel to 70 degrees. As you weld on the T-fittings, quickly check them against the sliding adjustable bevel

Fabricate the bottom assemblies for the center arch by welding two T-fittings to a 2-in. (5.1cm) piece of pipe. As with the corner assemblies, set one T-fitting at a 70-degree angle. The two assemblies should mirror each other.

Begin putting together the base of the pen by cutting a pair of side pieces at 31¼ in. (79.4cm). Note the positioning of the center joint assemblies. Use the roll of mesh to confirm that the space will be just right for fastening the screening.

Complete the sides of the base by cutting a pair of pieces 33¼ in. (84.5cm) long and welding them to the center and corner assemblies.

Assemble each end of the base by welding a T-fitting to two 35-in. (88.9cm) pieces. Once assembled, use one as a guide for the center crosspiece.

Continued

Making a PVC Hen Pen (cont'd)

12

Finish building the base by welding together the components made in steps 9, 10, and 11. Make sure that the T-fittings in the end pieces are upright. A scrap of 2x4 makes a handy guide. Important: let all of the assemblies you have made thus far dry for at least 2 hrs.

13

Begin each arch by welding one end to the base. For the end arches, line up the upright member with the frame as shown. For the center arch, make sure the cross is lined up with the base when you weld it in place.

14

Bend and weld the end arches. This is the tough part. You need to prepare your joints and assemble them quickly before the cement dries. The nature of the structure is such that you won't be able to give the joint a twist as you assemble—just push.

15

Complete the frame by adding two 34-in. (86.4cm) pieces to the top. The frame will give enough for you to fit them in and twist them while making the weld.

16

Make a swinging gate at one end by cutting a piece of pipe an inch shorter than the height of the arch. Mount it from the top of the arch so that it swings on the 2½-in. (64mm) bolt. Fasten the bolt with washers and two nuts.

Continued

So You Made a Mistake

It is easy for something to go wrong while assembling this project. By having a few straight fittings on hand, you can cut apart your work, make necessary adjustments, and reassemble. The result will look almost intentional.

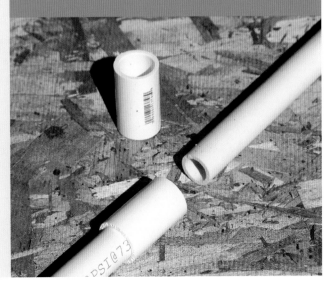

Making a PVC Hen Pen (cont'd)

17

Add the plastic hardware mesh to the pen, beginning at the top and rolling it down one side. Lace nylon ties through the mesh and around the pipe every 6–8 in. (15.2–20.3cm). Trim the mesh using aviation shears or heavy scissors (inset).

18

Cut a tarp to cover one-half of the pen—or whatever area you choose for sheltering your hens from sun and rain. Work out the size of the tarp by laying it on the pen and allowing for the bungee cords. Use a grommet punch and die to add grommets where you need them.

19

Create a door by attaching the mesh to the swinging pole and the side of the arch. Do not attach the mesh to the bottom of the frame. Trim the cable ties.

20

Add a bolt to the left side (right side looking in) of the pen to hold the door open when chickens enter or exit. Add another bolt to secure the door when you close it.

Making a PVC Hurdle

Originally made by British shepherds, a hurdle is a portable fence panel that can come in awfully handy when corralling livestock. The ancient version was made of split wood. (See "The Ancient and Original Hurdle," page 121). This PVC variant is light and storable, ideal for herding poultry.

While assembling, you'll find it much easier to line up the fittings if you have a large, flat work surface. Getting the elbow, T-, and cross-fittings oriented correctly is the only challenging aspect of this project. The hurdle shown uses almost all of two 10-foot pieces of ½-in. (1.3cm) PVC. You can make a larger hurdle, but be aware that the longer it gets, the more the PVC will flex.

Before beginning, see "The Method," on page 112, for the basics on welding PVC.

Tools	Materials
Measuring tape	2 10-ft. (3m) pieces of ½-in. (12mm) PVC pipe
Black felt-tip marker	2 ½-in. (12mm) T-fittings
Fine-tooth saw	4 ½-in. (12mm) 90-degree elbows
Miter box	1 ½-in. (12mm) cross-fittings
Coarse sandpaper	2 ½-in. (12mm) caps
Framing square	PVC primer (clear, not purple, if available)
Aviation shears or heavy scissors	PVC cement
Nippers	1 3 x 4-ft. (91.4–121.9cm) piece of ½-in. (12mm) plastic hardware mesh
	20 8-in. (20.3cm) cable ties

Light and compact, this PVC hurdle is a new variation on an ancient theme—and ideal for corralling poultry.

PVC Hurdle Exploded View

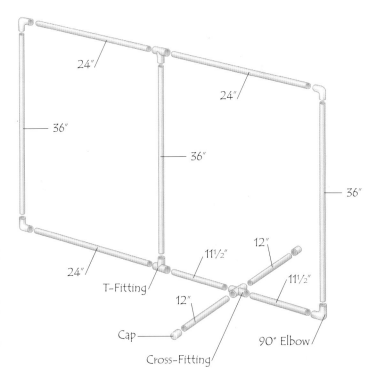

24"
24"
36"
36"
36"
24"
11½"
12"
12"
11½"
T-Fitting
Cap
Cross-Fitting
90° Elbow

Making a PVC Hurdle

1

Cut three 36-in. (91.4cm) pieces of ½-in. (12mm) PVC using a miter box and a fine-tooth saw. Weld 90-degree elbows to each end of two pieces. On the remaining piece, weld a T-fitting at each end. In each case, orient the fittings in exactly the same direction.

2

Cut two 11½-in. (29.2cm) pieces of ½-in. (12mm) pipe. Weld them to opposite sides of a cross-fitting.

3

Cut and weld two 12-in. (30.5cm) pieces to the open side the cross-fitting. Cap them to make the legs of the hurdle.

4

Weld the 36-in. (91.4cm) piece with the T-fittings to one of the 11½-in. (29.2cm) pieces attached to the cross. As you weld it, quickly square up the joint using a framing square. If you make a mistake, cut away the piece and use straight connectors to make a fresh start. (See page 117.)

5

Lay the project on its side, and add the remaining pieces. Laying the hurdle flat ensures that it will be straight and its legs will be at 90-degree angles.

The Ancient and Original Hurdle

Made by weaving long lengths of split hazel wood around upright stakes, hurdles were once used by the thousands to pen sheep. Making a solid and long-lasting hurdle is an art, but it's worth trying if you have plenty of hazel, willow, or other kind of straight, supple wood.

6

Roll out ½-in. (12mm) plastic hardware mesh. Use a scrap of 2x4 to keep it from rolling back on you. Attach the mesh using an 8-in. (20.3cm) cable tie every 12 in. (30.5cm) or so. Cut off loose ends using nippers.

CHAPTER 3

Housing Chickens

LAYING HENS ARE IDEAL for backyard homesteading. They require minimal space, are quiet, produce little odor if their run is kept clean, and are wonderfully productive. And in an increasing number of cities and towns, they are legal.

If you are considering building a chicken coop, your first task will be determining how many chickens you are allowed and any restrictions on their dwellings. Most municipalities do not permit you to own roosters, but they do allow three or four laying hens; therefore, smaller coops get detailed attention in this chapter. Siting is also an important consideration. Towns require that a coop be a certain distance from your neighbors' lot lines—anywhere from 25 to 100 ft. (7.6–30.5m).

The Human Factor

It is pretty clear what chickens need. (See sidebar below.) But chicken keepers have needs, too, so include the human factor in your coop planning. One primary goal should be getting the labor-intensive section of the coop off the ground. If you have to get down on your knees in the morning dew to replenish food, water, and litter—as well as gather eggs—the chores may not get done as often as they should. With the coop a couple of feet off the ground, nurturing the ladies is a pleasure. If the necessities—food, grit, litter, and oyster shell—are near at hand, all the better.

Choosing Frame Materials

Walls made of 2x4 framing with ½-inch plywood and a layer of siding make sense for human habitations, but they just are not necessary for chickens. Downgrading your specs makes sense. In moderate climates where insulation is optional, for example, a plywood box coop is fine. (See page 125.) Even if you want a cavity in the wall for insulation, 1x4 framing will do. Lighter construction can save you money—and a lot of aggravation if you ever want to relocate your coop.

So why do people immediately reach for 2x4s? One good reason is that they are easy to work with. If your nailing accuracy leaves a lot to be desired, a 2x4 offers a nice, big target. And if your engineering is a little off, the flaws will show up more readily in light construction. In sum, overbuilding is just plain easier and probably the way to go if you are new to carpentry. Assess your skills and frame accordingly.

What Chickens Need

- **Privacy:** 12 x 12 x 12-in. (30.5 x 30.5 x 30.5cm) nesting boxes
- **Room to perch:** 8 in. (20.3cm) of perch per bird
- **Feed access:** 4 in.(10.2cm) of feed trough per bird
- **Room to roam:** 4 sq. ft. (1.2 sq. m) of indoor/outdoor space per chicken
- **Security:** a coop with a closable door
- **Shelter:** a rain- and wind-proof coop
- **Ventilation:** a means of cooling the coop in hot weather
- **Clean water and feed:** suspended containers of water and food
- **Foraging space:** an outdoor run
- **Clean litter:** easy access for replacing litter

124 Building a Chicken Coop and Run

142 Building an A-Frame Chicken Tractor

152 Coping with the Cold

156 Prepping for Extreme Heat

Building a Chicken Coop and Run

If you plan a flock of three or four layers and want a portable coop with built-in feed storage, here is an option that you can build in a few weekends. Simplicity is the byword for this coop-and-run combo. Nesting boxes are built in instead of bumped out. No doors and hatches have to be fabricated. An optional storage area is included to keep feed, oyster shell, grit, and litter convenient.

This coop design is also easy on material expense. You need only three 4 x 8-ft. (1.2 x 2.4m) sheets of exterior plywood for the nesting area. It is self-contained—a plywood box that provides nighttime security for your hens. The 2x4 frame for the run is the exoskeleton for the project. If you can find them, 2x3s do the job as well and will cut the weight. If you want a larger coop, increase it to 4 x 4 ft. (1.2 x 1.2m) and adjust the run dimensions accordingly.

This project requires experience using a circular saw to make long, straight cuts and accurate plunge cuts. (See Step 4 on page 129.) If you have doubts about your skill, get some scrap plywood and try out the techniques shown. They are not hard to do but take a bit of practice. Always wear eye and ear protection when sawing.

The coop suits most climates. An air space between the corrugated plastic roof and the plywood roof of the coop buffers the heat of the sun in the summer and insulates somewhat from snow and ice in the winter. In very cold climates you may want to add insulation to the coop and consider metal corrugated roofing to handle snow loads. If you prefer more coop space or want to add an interior roost, eliminate the feed-storage area.

Neat and complete, this coop combines feed storage, nesting boxes, and safe harbor for the night. The screened run offers almost 24 sq. ft. (7.3 square meters) of grazing space. And this coop-and-run combo is portable. As the layout of your backyard homestead changes, you can drag the coop to another spot.

When it comes time to clean out the coop (twice a week being ideal), a hook on the ventilation hatch, top, holds the cleanout door open for easy access. The height of the coop floor is just right for getting used litter into a wheelbarrow.

Gathering eggs is easy with this fold-down hatch, above, and so is cleaning out the nesting boxes. Both the boxes and the coop floor lift out to make serious cleaning a breeze when needed.

In hot weather, the ventilation hatch aids air circulation, right. If a surprise rainstorm hits, the slant of the hatch will protect the interior.

Plywood Cutting Guide

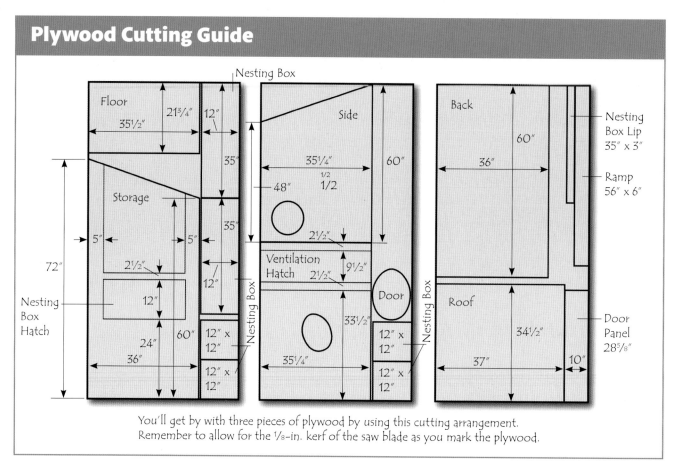

You'll get by with three pieces of plywood by using this cutting arrangement. Remember to allow for the 1/8-in. kerf of the saw blade as you mark the plywood.

Tools

Circular saw with plywood blade and a 12–24-tooth carbide-tip blade
Saber saw
Sawhorses
Measuring tape
Cordless drill-driver (having two is ideal)
Drill and driver bits
Pry bar
Hammer
Handsaw
Drywall T-square
Speed square
Framing square
Squeeze clamps
Sliding adjust-able bevel (optional)
Compass
Rasp
1- and 1½-in. (25 and 38mm) hole-cutting saws
Aviation shears

Materials

3 sheets of exterior-grade 4 x 8-ft. (1.2 x 1.4m) ½- or ⅝-in. (1.3 or 1.6cm) plywood
6 8-ft. pressure-treated 2x4s
4 12-ft. (3.7m) pressure-treated 2x4s
2 8-ft. (2.4m) pressure-treated 2x2s
2 8-ft. (2.4m) 2x2s
1 8-ft. (2.4m) pine ½ x ¾-in. (1.3 x 1.9cm) parting stop
2 sheets of 2 x 8-ft. (2.4m) corrugated plastic roofing

1 4 x 25-ft. (1.2 x 7.6m) roll of ½ x ½-in. (1.3 x 1.3cm) plastic screening
Exterior glue
4d cement-coated nails
1¼-in. (32mm) exterior screws
2-in. (51mm) exterior screws
3-in. (76mm) exterior screws
100 #12 x 1-in. (25mm) roofing screws
Wire staples
1-in. (25mm) brads
4 2-in. (51mm) barrel bolts
8 2½-in. (63mm) hinges

2 4-in. (102mm) hinges
6 flush-type wire rope clamps
12 x 12-in. (30.5 x 30.5cm) acrylic sheet
4 ¾-in. (19mm) pan-head screws
14 ft. ³⁄₁₆-in. (5mm) wire rope
3 hooks and 3 eyes
2 large hooks for feeder and water dispenser
1 8-ft. (2.4m) ¾-in.-dia. (19mm) conduit
Pieces of tree limbs for roosts

Coop Front View

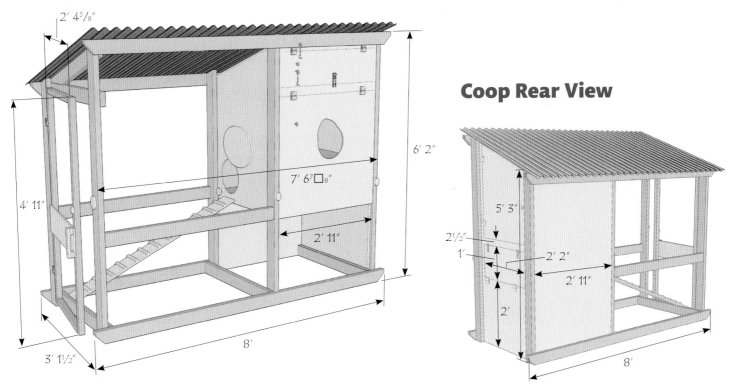

Coop Rear View

Coop Exploded View

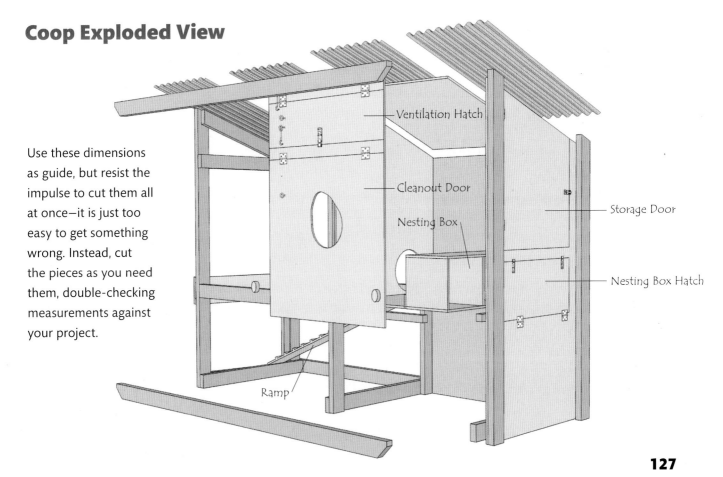

Use these dimensions as guide, but resist the impulse to cut them all at once—it is just too easy to get something wrong. Instead, cut the pieces as you need them, double-checking measurements against your project.

Ventilation Hatch

Cleanout Door

Nesting Box

Storage Door

Nesting Box Hatch

Ramp

Building a Chicken Coop and Run

Set up a work area. This job goes much easier if you have a couple of sawhorses on which you can set two 8-ft. (2.4m) 2x4s and a piece of plywood or oriented-strand board (OSB) as a flat, stable base for laying out and cutting your plywood.

Mark the plywood, following the cutting diagram on page 126. Mark for only one side at a time, including any outlines for doors and hatches. To cut the incline for the roof, strike a perpendicular line even with the lowest point of the roof, measure up 12 in. (30.5cm), and use a straightedge to mark the incline. Under the plywood, place scrap 2x4s positioned to support the sheet as you cut it.

3

Attach the hinges and catches before cutting, and you'll save yourself a lot of trouble later. Then remove the hardware; make the cutouts; and reattach the hardware. You'll find you've achieved a near-perfect gap on all sides of the cutout.

4

To make a plunge cut, position the blade over the cut line. Rest the front of the base, or shoe, of the saw on the plywood. Be sure that the blade is not touching the plywood, and then start the saw. Carefully lower the blade, watching to see that it is on the cut line. Once the saw is fully on the plywood, push it forward to the corner marks.

5

Complete the cutout. You'll notice that the cutouts will remain suspended in place. That's because the circular shape of the blade keeps it from cutting all the way to the corner. And here the fun begins. Replace all the hardware. Then use a handsaw or saber saw (inset) to carefully complete the corner cuts.

Continued

Building a Chicken Coop and Run (cont'd)

Consider the Pneumatic Option

Hand-nailing plywood corners is time consuming and can lead to splits. A pneumatic nail gun is faster and does the job with nary a crack. It is useful for fabricating feeders, hives, and mangers—all of the smallish projects that a backyard homestead requires. You'll also find the compressor handy for things like topping up tire pressure on wheelbarrows. One caution: used with glue, pneumatic fasteners work well after the glue sets, but screws and cement-coated nails hold better immediately. A basic kit, including pancake-type compressor, hose, and nail gun (shown), is the least-expensive option at about $300. Renting a similar setup is about $70 a day.

Cut out the entire side. Begin by using a speed square (inset). Push the saw forward until it is fully on the plywood. Leaving the saw in place, clamp the drywall T-square in position. Pull the saw back, start the saw, and complete the cut.

Cut the coop opening. Use a 5-gal. bucket (inset) as a tracing guide for marking the entryway for the hens. Bore a ½-in. (12mm) access hole, and then use a saber saw to cut the opening. Complete the other sides, cutting out any necessary hatches. Remember that opposite sides should be the same width—double-check as you measure.

8 **Begin assembling the coop.** To do it solo (as shown), tack a couple of 2x4s to your worktable and, with a scrap of wood clamped to one coop side to help you set up the overlap, glue and nail the pieces to together. Begin with the longest sides—the back and the nesting box/storage side. Use exterior glue and 4d cement-coated nails.

9 **To add the remaining sides,** attach temporary legs so that you can position the sides while fastening. Pay attention to which sides overlap. For the front side, add 2x2 stiffeners behind the 2½-in. (64mm) strips to which the hinges are attached.

Continued

Add Anti-Rot Protection

The edges of even exterior plywood are susceptible to decay from moisture. Without protection, exposed edges will eventually delaminate and rot. After sealing the edge with wood preservative, add a strip of plastic J-channel used for drywall. Cut it to length; run a generous bead of silicone caulk inside; and push it onto the plywood edge.

Building a Chicken Coop and Run (cont'd)

10

Add the roof. It protects the hens from the heat of the sun in the summer and captures insulating air beneath the corrugated plastic roofing in colder weather. Cut the piece so that it sets down into the box. Do a test fitting before gluing and nailing. Tacking nails at the four corners about ½ in. (1.3cm) down from the top edges will hold the roof panel in place while you fasten it (inset).

11

Install the floor and nesting boxes by first adding 2x2s to the interior of the coop to hold the flooring and boxes. Cut the floor, and do a test fit. Don't fasten the floor and boxes in place. Doing a radical cleaning of the coop is easier if you can pop them out.

Build Nesting Boxes

1. Cut and mark front and back pieces for the nesting box: cut two 12 x 35-in. (30.5 x 90cm) pieces of plywood. Position the pieces side by side, and using a framing square, mark the location of the sidepieces.

2. Cut four 12 x 12-in. (30.5 x 30.5cm) sides, and start the nails in the front and back: flip one long piece over, and mark it for the nails as shown. Tack all of the nails before assembling the boxes. Drill pilot holes along the edges.

3. Glue and nail the nesting box. Enlist a helper if possible to help assemble the box, or use a spring clamp (shown) to help hold the sides upright as you glue and nail them. A pneumatic nail gun (inset) makes quick work of this job.

12

Measure for the frame uprights so that they extend 1¾ in. (4.5cm) above the coop roof. Use a scrap from one of the inclined sides to mark the angle, or use a sliding adjustable bevel (inset) to capture and transfer the angle. **Continued**

Free Coop Plans

Go to following sites to search out free coop plans. Also consider looking at galleries in some of the chicken-keeper community sites, such as backyardchickens.com. Often, owners are pleased and proud to provide information about how to replicate their coops.

www.poultry.purinamills.com
www.motherearthnews.com
www.pvcplans.com
www.backyardchickens.com

Building a Chicken Coop and Run (cont'd)

13

Cut three uprights for the front of the coop and three for the back. Cut each set of three to the same length and angle. Wear a face mask in addition to eye and ear protection when cutting pressure-treated lumber.

14

Cut the horizontal framing members. Choose four straight 8-ft. (2.4m) 2x4s for the horizontal framing. Measure 1 in. (2.5cm) from the top of each, and mark for a 45-degree cut.

15

Assemble the frame using 3-in. (76mm) deck screws. Fasten each corner with one screw, and then square it up using a framing square. To keep the frame from going out of square, use a scrap of lumber as a cross brace on one of the upper corners (inset). Add a second screw to each corner. Do not add the middle upright at this point.

16

Attach the frame to the coop using 1½-in. (38mm) deck screws fastened from inside the box. Add the center uprights and cross members, again fastening from inside the box (inset). Use 3-in. (76mm) deck screws where you join two 2x4s.

17

Build the run access door. Measure the opening, and subtract ½ in. (1.3cm) for ¼-in. (6.4mm) clearance on both sides. When opened, the door needs to clear the ground by 1½ in. (3.8cm). Drill pilot holes, and fasten two 3-in. (76mm) deck screws in each corner.

18

Square up the door frame, and add the center support panel to keep the door from sagging. Fasten it using two 1¼-in. (32mm) screws on each side.

Continued

Building a Chicken Coop and Run (cont'd)

19

Position the door, and clamp it in place so that there is about a ¼-in. (6.4mm) gap on each side. Clamp a speed square in place to ease lining up the hinges. Fasten both hinges.

20

Add the door latch. As an alternative to a hook and eye, use a hole-cutting saw to make a 1½-in. (3.8cm) bumper for the latch. Freehand-sketch a 3 x 1½-in. (7.6 x 3.8cm) egg-shaped latch.

21

Cut out the egg latch using a saber saw. Use 1½-in. (38mm) exterior screws and washers to attach the bumper and latch.

22

Trace and cut out additional latches. Once you've got one good egg, make two more for the cleanout door (shown). You may also want to use them as a substitute for some of the hardware latches.

23

Make a door handle by cutting a 7-in. (17.8cm)-long piece of 1x2 and two 1½ x 1½-in. (3.8 x 3.8cm) blocks.

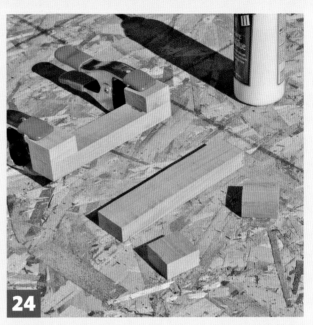

24

Apply exterior glue, and clamp the pieces together. When the glue has dried, drill holes at each end. Set the handles aside for painting.

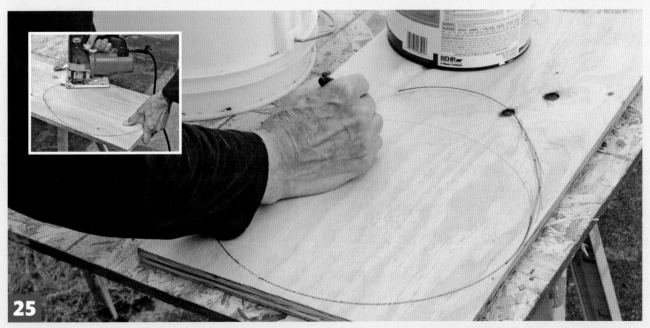

25

To mark an egg-shape coop door, begin by tracing around the top, not the bottom, of a 5-gal. bucket. Position a 1-gal. paint can so that it overlaps the tracing by about 4 in. (10.2cm), and trace around it. Now exercise a little artistry: sketch gentle curves to join the two circles and get a pleasing egg shape. Use a saber saw (inset) to cut out the door.

Continued

Building a Chicken Coop and Run (cont'd)

26

Finesse the egg-shape coop door to achieve a pleasing silhouette. An egg's perfect shape is surprisingly difficult to manage. To smooth out any imperfections, use a rasp (shown) or a belt sander if you have one.

27

Attach the door using a 1½-in. (38mm) exterior screw (and washers front and back) 2 in. (5.1cm) off center so that you can open it using a vertical cable (inset). Cut 1- and 1½-in. (2.5 and 3.8cm) discs; add two washers to the 1-in. (25mm) disc; then attach them using a 1½-in. (38mm) exterior screw and washer.

28

For the ramp, cut plywood 56 in. long and 6 in. wide. It needs to rest under the coop opening so that the door can swing closed. At the opposite end, it rests on the horizontal framing by the door. Cut ¾ x ¾-in. (1.9 x 1.9cm) stock 6 in. (15.2cm); glue and nail one every 6 in. (15.2cm). Add two 2½-in. (64mm) screws beneath the lower ramp end so you can remove it for cleaning.

To Paint or Not to Paint

If you choose the unpainted look (below), seal the plywood immediately. The pressure-treated framing will withstand decades of weather without being sealed. If you choose to paint the coop, use exterior semigloss paint, which is easy to clean. If you prefer painting with a variety of colors, dismantle the doors and hatches for a neat job (far right). Warning: paint both sides of these pieces so that they won't warp. Try to use low-VOC (volatile-organic-compound) paints and sealers.

29

Cut the ½ x ½-in. 1.2m screening. The framing will take 4-ft.- and 2-ft. (1.2m and 61cm)-wide sections. Screening is challenging to cut because it tends to spring on you. Wear gloves and, using aviation shears, cut a section to length. Turn it over, and "back roll" it so that it flattens. Notch it as necessary to fit behind the upper and lower horizontal

30

Fasten the screening using #12 x 1-in. (25mm) roofing screws. The beauty of roofing screws is that they are self-tapping and therefore won't split the wood. In addition, they can be backed out if you need to make an adjustment. Buy about 100—you will use them on the roof as well. They will require a ⁵⁄₁₆-in.(8mm) hex bit for your drill-driver.

Continued

Building a Chicken Coop and Run (cont'd)

31

Cut the corrugated plastic roofing with aviation shears. Using a framing square as guide, mark the midpoint of an 8-ft. (2.4m) piece of corrugated plastic roofing. If you anticipate heavy snow load, you may need to use metal roofing, which will require cutting with a circular saw and carbide-tip blade.

32

Fasten the roofing using #12 x 1-in. (25mm) roofing screws. Apply one fastener every three or four corrugations. By overlapping the sheets one or two corrugations, you'll be able to completely cover the coop.

33

Attach an acrylic sheet to the egg window. Before removing the protective plastic from the acrylic sheet, cut it to size using a utility knife and straightedge so that it extends at least 2 in. (5.1cm) beyond the opening as shown. Make several scores, and then hold it over the edge of your work surface. Firmly hold a piece of scrap lumber along the cut line, and crack off the unneeded piece. Predrill holes for four ¾-in. (19mm) pan-head screws.

Finishing Touches

While the frugal farmer may settle for a bent nail as a hook here and a length of baling twine as a hanger there, the following bits of hardware will make chores slightly easier and give decades of service.

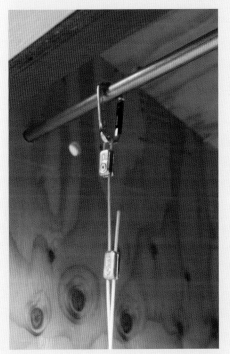

A simple hook attached to the ventilation hatch and an eye attached to the cleanout door hold the door open when it is time to remove the litter or replenish the feed, above.

Electrical conduit pipe is an inexpensive, strong support for water dispensers and feeders, left. A ¾-in. (19mm)-dia. hole in the framing and the coop supports it.

A cable looped around a screw and washer attached to the door allows you to open and close the coop door without entering the coop, right. A screw eye up top guides the cable. A hook and eye on the front of the coop holds the door open (inset).

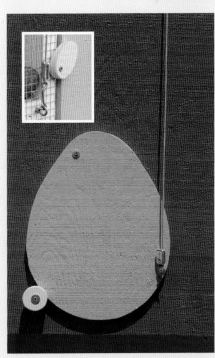

Building an A-Frame Chicken Tractor

A chicken tractor allows you to let your hens graze in a controlled fashion so that they can eat plenty of grass, insects, and grubs without turning your yard into a moonscape. That is because you can easily move them along, coop and all, to fresh territory after a day's grazing.

This tractor offers all the amenities—shelter, nesting box, ventilation, feed, and water—in one mobile package. The 6 x 8-ft. (1.8 x 2.4m) tractor comfortably holds up to four layers, either as a permanent home or as a mobile shelter for summertime excursions.

It is also people friendly. The tiller-like handle makes it easy to lift and steer. The rear hatch allows access to the nesting box, which lifts out for easy cleaning. The hinged vent helps cool the coop on hot days. The door to the run opens wide for herding hens into

Whether it is permanent or just a seasonal home for your chickens, this tractor is light enough for you to easily move it to greener pastures every day or so. Coop access includes an optional vent (inset) to cope with summer heat.

the tractor. And the tractor is light, thanks to the inherent strength of its 2x2 triangular framing.

Two of the triangles form the frame for the plywood coop; a door is hinged to the third. The plywood coop keeps the whole thing rigid without fussy cross bracing. In fact, simple construction and affordable materials make this an ideal starter project. One caveat: the wheels are attached to a 2x4 with lag screws—an inexpensive and serviceable method, but one that may not withstand the rigors of uneven terrain. If you have rough pasture, consider larger wheels and a continuous steel axle.

Tools	Materials	
Circular saw with a 12–24-tooth carbide-tip blade	3 4 x 8-ft. (1.2 x 2.4m) sheets of ½-in. exterior plywood	Wire staples
		1½-in. (38mm) exterior screws
Saber saw	6 12-ft. (3.7m) pressure-treated 2x2s	1 6-ft. (1.8m) cedar fence board
Sawhorses		1 4-ft. (1.2m) 1x1
Measuring tape	6 8-ft. (2.4m) pressure-treated 2x2s	2 wheels
Cordless drill-driver (having two is handy)	1 10-ft. (3m) ⁵/₄ x4 pressure-treated decking	2 5-in. (127mm) lag screws with washers to fit the wheels
Drill and driver bits		Paint or sealant
Handsaw		2 strap hinges
Drywall T-square or framing square	2 8-ft. (2.4m) ⁵/₄ x4 pressure-treated decking	2 gate hinges
Speed square		2 small hinges
Clamps	48-in. (1.2m)-wide ½ x ½-in. (1.3 x 1.3cm) screening	4 hook and eyes
Block plane or rasp		¹/₈-in. (3mm) hardboard scraps as spacers
Adjustable wrench	60 #12 x 1-in. (25mm) roofing screws	4 ft. (1.2m) of 4–6-in. (10.2–15.2cm)-wide aluminum flashing
Hammer	2½-in. (64mm) exterior screws	
Aviation shears		

Chicken Tractor Elevations

Begin by assembling three 6 x 6 x 6-ft. (1.8 x 1.8 x 1.8m) frames. Next, add the 4 x 4-ft. (1.2 x 1.2m) ½-in. (12mm) plywood sides of the coop, then the bottom. Use the structure itself to mark the plywood triangles that form the front and back of the coop. The tractor has two doors, one for access to the run, another for gathering eggs and cleaning out the coop. Attach a 10-ft. (3m)-long piece of decking at a convenient height for lifting and steering the tractor.

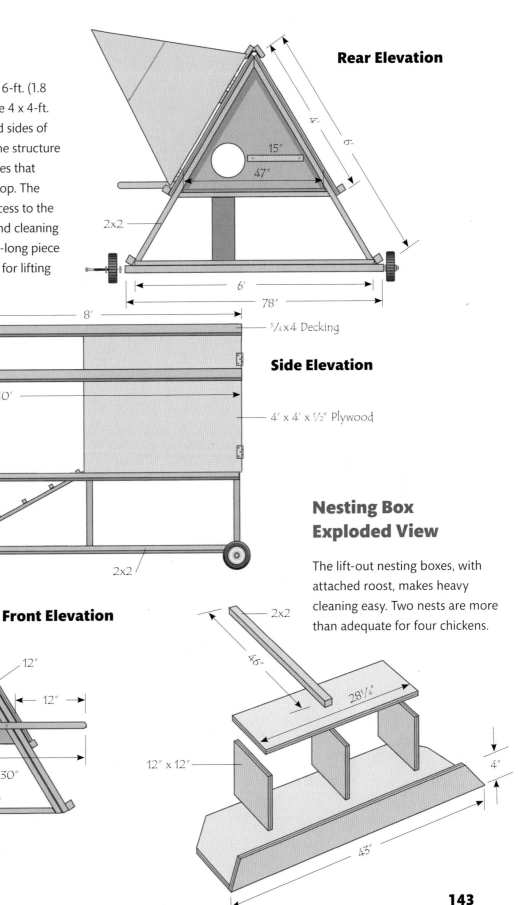

Rear Elevation

15"

47"

2x2

4'

6'

6'

78"

8'

⁵/₄x4 Decking

Side Elevation

10'

4' x 4' x ½" Plywood

6'

2x2

Nesting Box Exploded View

The lift-out nesting boxes, with attached roost, makes heavy cleaning easy. Two nests are more than adequate for four chickens.

Front Elevation

12"

12"

46"

30"

2x2

46"

28¼"

12" x 12"

4"

43"

143

Building an A-Frame Chicken Tractor

1

Cut nine pieces of 2x2s 6 ft. (1.8m) long. Form a triangle by clamping two of the corners while lining up the third. As you clamp, position the ends of the 2x2s so that they don't protrude beyond the edge of the piece they are overlapping (inset).

2

Mark for the angled cuts that will form the peak of the A-frame by overlapping the 2x2s and using a speed square or other straightedge as a guide.

3

Make the angled cuts using a circular saw. Extend the saw blade as fully as possible. You'll cut completely through the top piece. Use the partial cut in the lower piece as a cutting guide (inset).

4

Assemble three identical frame triangles. To avoid splits, position each screw away from the edges of the wood, angling the pilot hole. Use two 2½-in. (76mm) screws for the top and one each for the bottom joints.

5

Cut a 4 x 8-ft. (1.2 x 2.4m) piece of ½-in. (12mm) ply-wood in half, and position a frame, top down, ½ in. in from the edge of the plywood. Fasten the triangular frame using four 1½-in. (38mm) screws as shown; then attach a second frame similarly along the opposite edge.

6

Fasten the second roof piece even with the peak of the frame. If you can't enlist a helper, clamp a roughly 4-ft. (1.2m)-long scrap to stabilize the triangular frame as you install the plywood roof piece.

7

Cut the bottom piece 48 x 47 in. (121.9 x 119.4cm). Hold it in place to mark notches to allow for the 2x2 framing. The notches need to be 1¾ in. (4.4cm) long to allow for the slant of the 2x2s. Start the notch with a circular saw; finish using a handsaw as shown.

8

Fasten the coop bottom in place using 1½-in. (38mm) exterior screws. Drill pilot holes to avoid split-ting, angling the holes so that the screws firmly hold the pieces together. Join the bottom piece to the roof pieces using screws every 8–10 in. (20.3–25.4cm).

Continued

Building an A-Frame Chicken Tractor (cont'd)

9

Mark and cut braces to fit beneath the floor. Line up the piece, and mark it in place as shown. Mark one end and trial-fit it. Adjust the cut as needed before marking and cutting the other end.

10

Fasten the braces to the frame using angled 2½-in. (63mm) screws. Clamping the piece in place makes the job easier. Attach both ends to the framing, and then fasten the plywood floor to the brace using 1½-in. (38mm) screws every 8–10 in. (20.3–25.4cm).

11

Mark and cut the triangular front and back pieces by cutting a 4 x 8-ft. (1.2 x 2.4m) piece of plywood in half and clamping one piece in place. Mark along the roof; flip the piece; and cut. Position it again; mark for the bottom; and cut (inset) so that it lines up with the roof pieces. Repeat for the other half.

12

Cut a hole in the front piece for the chicken access opening. Use the bottom of a 5-gal. bucket to mark the 14-in. (35.6cm)-dia. hole. Make sure the bottom of the oprning is about 4 in. (10.2cm) above the floor. Bore an access hole, and use a saber saw to cut the opening.

13

Add a vent to the back piece (the access door) by marking a line about 12 in. (30.5cm) down from the peak. Cut the vent; line up the pieces; and install small hinges. Just below the cut line, add the pivot bar to hold the vent closed when not in use.

14

Add strap hinges to the access door by first setting ⅛-in. (3mm) spacers to provide a gap. Attach the hinges. You may need to bore additional holes in the hinge to avoid fastening too close to the edge of the door.

15

Build the run door in the same way as the A-frames. To size it, assemble it within the remaining frame using ⅛-in. (3mm) spacers. The door bottom must be at least 1 in. above the frame bottom to clear vegetation as it swings. Use two 2½-in. (63mm) screws at each joint.

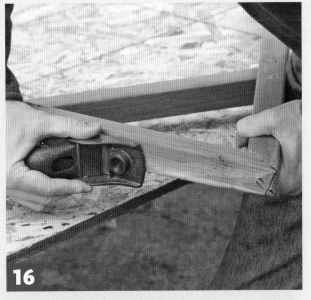

16

Plane or saw the back of the topmost joint so that the door will clear the framing when you shut it—a quirk of triangular doors. You'll need to carve out a fairly deep area for clearance.

Continued

Building an A-Frame Chicken Tractor (cont'd)

17

Apply the hinges to the door, and then fasten it onto the frame. Test the swing; you may have to further plane the top edge started in Step 16 so that it clears the frame easily.

18

Attach the 8-ft. (2.4m) horizontal 2x2s to the bottom and midsection of the coop assembly using 2½-in. (63mm) screws.

19

Attach the door A-frame and door by drilling pilot holes and fastening the 2x2s to the door frame using 2½-in. (63mm) screws.

20

Add the crosspiece to the door. Begin by making the angle cut on the hinge side. Set the crosspiece in place, and mark a cut line on the other side so that an 8-in. (20.3cm) "handle" is left. Round the end. Drill pilot holes, and attach the crosspiece.

21

Set a piece of flashing along the roof peak. You will have to bend it to suit the angle of the peak. For a smooth job, clamp a piece of 1-by to the flashing midway along its length and bend the flashing on a flat surface.

22

Attach two 8-ft. pieces of ⅝x4 decking, one on each side of the peak. Drill pilot holes through the flashing and decking, and fasten into the 2x2 A-frames using 2½-in. (63mm) screws.

23

Round the end of a 10-ft. (10m)-long ⅝x4 piece of decking, and attach it to the left side of the tractor (right side looking in). This is the lifting and steering bar for the project, so place it at a height that is comfortable for you.

Hen Herding

Chickens appreciate a ramble now and then. Letting them out to forage in the backyard is a joy to behold. Getting them back in the coop is another thing entirely. The girls can be wily, requiring at least two adults to herd them back home.

Here is a trick that saves anguish all around. Turn on the garden hose, and shoot a concentrated spray up into the air to create mini-rainstorms behind the chickens. They hate rain and will suddenly decide that the shelter of the coop looks very good indeed. With practice, you will be able to herd them into the gate with three or four squirts.

Continued

Building an A-Frame Chicken Tractor (cont'd)

24

Build the nesting box using the method shown on page 133. Keep the fit loose so that the box is easy to remove when cleaning out the coop.

25

Add a 2x2 roosting bar to the top of the nesting box. Attach a support for the roost on the front wall of the coop (inset).

27

Attach the 2x4 wheel support beneath the coop end of the tractor using two 2½-in. (63mm) screws where the A-frame overlaps the support. Note: If you have rough terrain, invest in larger wheels and a steel axle.

28

Apply the ½ x ½-in. hardware cloth using #12 x 1-in. (25mm) roofing screws or fencing staples. Install the ramp with the cleats attached. The ramp shown is made from a 6-ft. (1.8m)-long cedar fence plank.

26

Use lag screws (5 in. (127mm) long and just thick enough so that the wheel spins freely) and washers on each side of the wheels to attach them. Bore a hole about ¹⁄₈ in. (3mm) smaller than the shank diameter of the bolt into each end of an 8-ft. (2.4m) 2x4.

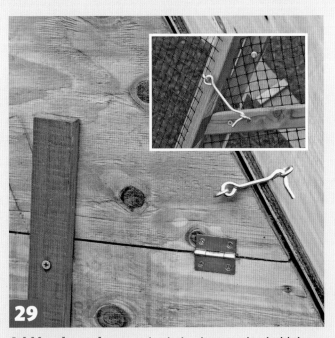

29

Add hooks and eyes to latch the doors and to hold them open (inset) when needed.

Hauling Hens

How do you haul hens in a vehicle without leaving it unfit for human habitation? A cage may be your first impulse, but is too open to prevent messes. Instead, find a large cardboard box in which your hens can stand up without crowding. Cut an opening, and cover it with something that keeps the girls in but also allows plenty of ventilation. In this case, a toddler stairway safety gate fits the bill. Throw in a drop cloth to keep your hens from slipping around.

Final Touches

Seal or paint the plywood for a longer life. If you choose to seal the inside of the coop, allow plenty of drying and off-gassing time before introducing the hens. Pressure-treated lumber needs no sealing, though it does no harm to seal the ends.

If you face hot summers or a lot of rainy weather, consider wrapping the coop with corrugated plastic or fiberglass. Simply add another 2x2 halfway up each side of the coop, and fasten the corrugated roofing material. The air captured between the corrugate and plywood adds valuable insulation from the sun. (See page 156 for a similar arrangement.)

Coping with the Cold

Chickens are tough birds that, with a little help, can handle anything your local climate can dish out. Unfortunately, that now means more extremes than ever. Prepping your chicken house with the right stuff can keep the hens happy and the eggs coming.

Prepping for Extreme Cold

What does it take to give your hens a healthy winter home? First, forget about a heated coop. Chickens like consistency and will develop respiratory problems if they frequently go from fierce outdoor cold to a cozy interior. A better approach is to make sure the chicken house is oriented to block the wind. Seal up any drafty cracks or gaps. One step in particular seems counterintuitive. Your impulse might be to pack the nesting boxes with plenty of straw to make a cozy little bed for your hens. In fact, straw holds moisture, and moisture is one of the causes of frostbite. Instead, turn to sand for the floor of the coop and the nesting boxes. It dries out quickly, slowing the threat of frostbite.

Having completed the basics, insulating the nesting area is the next natural step. In a couple hours, you can have the job done. One warning: Adding rigid foam board requires an ample service door for the coop, something that is well worth having anyway for cleaning. Otherwise, you may have to split pieces and hinge them with tape to get them inside the coop.

Tools	Materials
Tape measure	Rigid foam board
Marker	Rigid foam board
Straight edge	adhesive
Clamps	Clear acrylic sheet
Drywall knife	Flexible plastic strips
Mini hack saw	
or drywall saw	
Caulk gun	

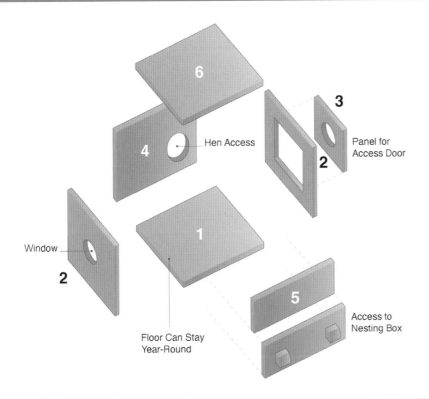

6

4 Hen Access

3

Panel for
Access Door

2

Window

2

1

Floor Can Stay
Year-Round

5

Access to
Nesting Box

Cut the rigid foam board panels so they pressure fit in place. Start with the bottom layer, covering the floor wall-to-wall. Then, fit the side walls with nesting box access and the coop entry for the hens. Fit the ceiling panel so it is supported by the side panels. Next, fit the other two wall panels. With all the panels positioned, you can mark for windows, nesting box access, and the coop entry.

Insulating a Chicken Coop

1

Measure and cut the panel for the floor, using a straight edge and a drywall knife for a clean cut. Install it foil-side up. There is no harm in keeping this piece in place all year long.

2

Cut out the opposite wall pieces so they fit snugly at the sides. Cut them short enough to allow for the ceiling panel, which will rest atop them. Chickens need all the daylight they can get in the winter, so mark the openings for any windows. Add double glazing with a piece of acrylic sheet, attaching it with rigid foam board adhesive. (See illustration above)

Continued

Insulating a Chicken Coop (cont'd)

3

Cut any openings using a mini hack saw or a drywall saw. Both let you do a plunge cut to begin sawing the insulation.

4

Cut a strip of something flexible to make a handle. Then, plunge cut a slot and glue in the loop with rigid foam board adhesive.

5

Cut the door access panel using an angled cut so it won't push through when set in place (see illustration, right). To remove it easily, install a plastic loop as you did for the nesting box panel.

Tip:

By cutting the insulation for the door panel at an angle, you can set the panel in place without it pushing through. Any windows can be double glazed with an acrylic sheet and rigid foam board adhesive.

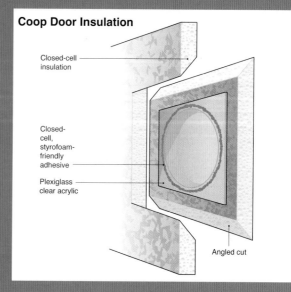

Coop Door Insulation

Closed-cell insulation

Closed-cell, styrofoam-friendly adhesive

Plexiglass clear acrylic

Angled cut

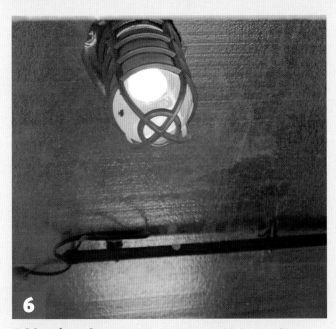

6

Add a piano heater and supplemental lighting after installing the wall and ceiling panels. The piano heater emanates just a few watts of heat, taking the edge off the cold without causing any respiratory problems.

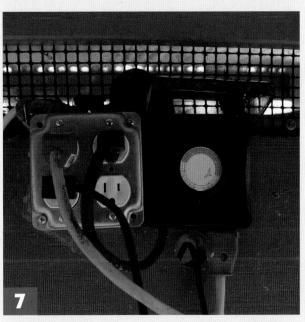

7

Install electrical outlets to power the piano heater (left) and heated waterer (below). (See pp. 262 for installing outdoor power.) If you supplement lighting in winter, you'll want to mount the timer close to your outlets.

8

Make a draft-free door by attaching strips of plastic over the nesting area entrance. After installing the door, tuck up a strip or two until the hens learn to push through the opening.

Food and Water

Make sure your feeder is capped off so snow and ice don't get inside. Hanging it so it is off the ground is good practice not only to keep it above the snow but also to keep hens from fouling it. Use a heated waterer anywhere temperatures fall below freezing.

Capped off feeder (left) and heated waterer (right) for the winter.

Prepping for Extreme Heat

In any temperature above 85 degrees, chickens can begin to overheat. You'll see them pant and spread their wings in an attempt to regulate their body temperature. They'll also lose their appetite. Heat can stress laying hens to the point where they'll stop producing eggs—and even die if they experience heat stroke.

In hot climates, misters are a must. Misters do not soak the chickens. Rather, they cool the air temperature by up to 20 degrees through evaporation. Next to shade, ventilation, and a consistent water supply, a misting system is your best safeguard for keeping chickens comfortable in extreme heat.

Tools	Materials
Heavy-duty scissors, cutting pliers, or utility knife	In-line misters Compatible tubing End cap for mister Zip ties

It won't take you more than an hour to install a misting system in your run. All it involves is access to water with a garden hose, mister heads, some appropriate tubing, and some zip ties. A variety of inexpensive kits designed as patio misters can be adapted to benefit your hens.

Cooling a Coop

That dark, reclusive area where chickens love to lay their eggs can become a hot box during excessively warm days. Here are some solutions to bear in mind when building your coop. First, remember that captive air insulates. If you can plan for an air space between the coop roof and the nesting area, you'll cut any transmittal of heat from the roofing. Second, a nice, chunky vent door at the top of the coop lets hot air escape. Finally, an elevated nesting area is not only handier for the egg gatherer, it allows air to circulate beneath the coop.

While chickens prefer dark, secluded areas to lay eggs, it's important to think of how heat is trapped in traditional coops. Ventilation is a must in extreme heat.

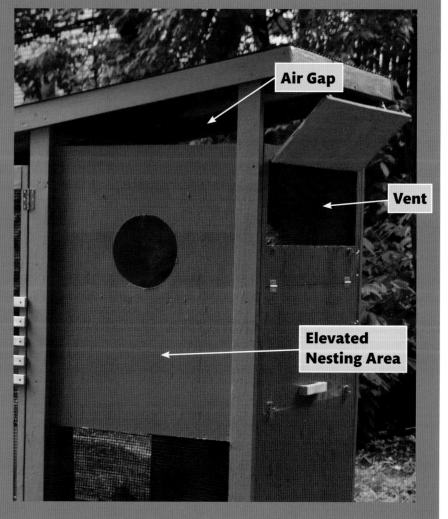

Air Gap

Vent

Elevated Nesting Area

Installing Coop Misters

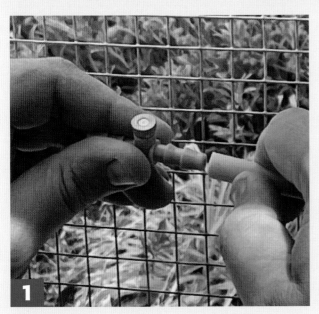

Install the first mister. About four feet (1.2m) above the ground, run the feeder tubing to your first mister. Cut the tubing with heavy-duty scissors, cutting pliers, or a utility knife. Push the mister into the tubing.

Attach the mister. Cut a piece of tubing about 1 foot (30.5cm) long, and push it onto the mister. Using a zip tie, attach the tubing to the run, making sure the misters point outward rather than down.

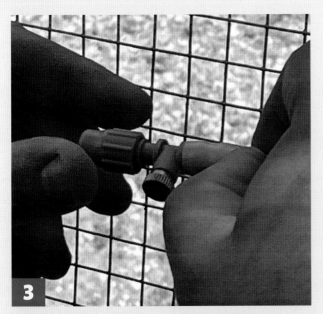

Cap off the last mister. In the same fashion, add more misters. One for every two or three chickens should do it. Zip tie each to the run. Add a terminal mister to cap off the system.

Attach the hose. Secure the feeder tubing to the run. Attach the garden hose. Consider adding an automatic water timer. Powered by a battery, these relatively inexpensive aids cue your misters to do their magic at only the hottest times of the day.

CHAPTER 4
Building Sheds

CHICKENS, GOATS, AND OTHER LIVESTOCK are surprisingly hardy, able to take rough weather and low temperatures in stride without supplemental heating or even insulation. But all animals need shelter from the wind and rain: a dry retreat away from the debilitating effects of hypothermia. That means a safe and solid shed. It need not be a paragon of carpentry perfection, just something strong and utilitarian.

Tools and supplies need similar protection. You must keep tools out of the weather, and you need to store them properly. (See page 211 for one option.) Feed needs to be dry and protected from vermin. Miscellaneous items like stakes, pots, fencing, and other materials need a place where you can keep them clean and readily available. And if you are serious about starting your own plants, you really should have a dedicated potting shed.

When planning your shed or sheds, start with the site. Even if you live in an unincorporated area, check the local zoning laws. These laws exist to keep your new shed from adversely affecting your neighbors' quality of life—and their real estate value. Zoning laws typically boil down to the following:

- **Setback** is the buffer between your shed and your neighbors' lot lines—typically 12 ft. (3.7m) but sometimes more.
- **Lot coverage** (sometimes referred to as Lot Coverage Ratio, or LCR) is the percentage of your lot occupied by structures and paving. Most zoning regulations require that at least one-half of the lot be left open.
- **Easements** are rights-of-way that protect pathways for utilities. Never put a shed in their way.

Think through how you'll get electricity and/or water to your shed, should you need them. Most municipalities will not allow overhead power lines because they are unsightly and can be brought down in a storm, which means that you will have to dig trenches, the depth varying from 18 to 40 in. (45.7 to 101.6cm), depending on your locale. You must bury a water line in a separate trench deep enough to avoid damage from frost heave.

Codes and Permits

Out in the country, building codes, as well as covenants and zoning regulations, may be less restrictive than in town. If you are building a shed on a 5- or 10-acre lot, local building codes and zoning may not impact your design at all. It is important to remember, however, that building codes are good guidance for best construction practices and minimum design standards. The town or county building department will tell you the code to which you must adhere. Increasingly, the International Building Code (IBC) is the standard that applies, but some counties, especially in hurricane- or tornado-prone areas, may demand that structures be beefed up beyond IBC requirements.

The requirements for building permits vary widely.

Some towns and cities exempt structures that are built on non-permanent foundations, such as concrete blocks or wood skids resting on gravel. If you must obtain a building permit, apply for one at your municipality's building department. The kind of permit you get and the fee you pay are usually related to the proposed square footage or cost of the shed. You may need to provide rudimentary building plans, and it may take a few days or a week for permit approval. If you are asking for a zoning variance, your permit may take even longer and be subject to public notice and hearings.

Think twice about forging ahead without a permit. In the worst case, you may be asked to remove the shed. For small structures, however, you can usually obtain a building permit after the fact and pay a fine.

160 Building Basics

168 Saltbox Shed

194 Goat Shed

206 Roofing Alternatives

210 Setting Up a Backyard-
Homestead Shop

Building Basics

Building a shed is a great way to hone your carpentry skills. Small cosmetic mistakes are unimportant. After all, this is farmyard-type carpentry intended primarily as shelters for animals and equipment. Your animals will not be bothered by a little "gapitis" in your measuring and cutting. But they will be bothered if the structure comes down on top of them. To avoid that and ensure that your shed will be an asset to your property, it is worth reviewing some building basics.

For floors, create a frame the size of the perimeter of the building using 2×4 or 2×6 joists positioned every 16 inches on center in the interior of the frame. For the sheds in this chapter, this kind of floor simply rests on foundation blocks. The weight of the structure keeps it in place. However, some municipalities require ground anchors to hold the shed on its foundation.

Floor Framing

1. Place the rim joists on top of the sill (with both slab and wall foundations) or the beam (on pier foundations for a small shed as shown here).

2. Check that the perimeter frame is square. Tack the joists in place, and using a framing square at the corners, roughly square it up. Then measure the overall diagonals, corner to opposite corner. They must be exactly equal. Adjust if necessary.

3. End-nail the rim joists and header joists together using three 16d common nails. Check that they are tight against the sill. You should predrill nailholes near the ends to prevent splitting.

4. Mark the joist layout on opposite header joists using 16-in. (40.6cm) on-center marks, depending on the span and load. Make duplicate layouts on both boards before installing the joists.

5. End-nail the joists through the header joist using three 16d cement-coated nails or 3-in. (76mm) deck screws. Local codes may require that you also set the joists into galvanized-metal joist hangers.

Building Walls

Stick-built walls are constructed using a modular system designed to provide a combination of structural strength and nailing surfaces for sheathing materials inside and out. Adding windows and doors usually requires extra studs because the openings rarely follow the exact modular framing layout.

A horizontal 2×4 (or 2×6 in some cases), called a soleplate, forms the base of the wall. Studs sit on the soleplate, generally every 16 in. (40.6cm) on center. Two horizontal 2×4s, called the top and cap plates, finish the wall.

There are three kinds of studs in most walls. Full-height studs, sometimes called king studs, run from the soleplate to the top plate. Jack studs, also called trimmer studs, run from the soleplate up alongside a full stud at rough openings. The top of the jack stud supports the header over a window or door. Short 2×4s called cripple studs fill in the spaces above and below a rough opening—from the soleplate to the sill of a window, for example.

They are spaced to maintain the modular layout and provide nailing surfaces for sheathing and surface materials.

Assembling the Wall. The most efficient way to build walls is to assemble the components on the floor and tip them into position. Start by marking the top plate and soleplate. Tack or clamp them side by side so that the marks line up exactly. Mark for the studs 16 or 24 in. (40.6 or 61cm) on center. Then mark for the jack and cripple studs that will support windows and doors. You can save yourself a lot of cutting by purchasing 2×4 studs that are precut to 91½ in. (232.4cm) rather than buying standard 8-footers. Precuts make frame walls that are exactly 8 ft. high, perfect for applying 4 × 8-ft. (1.2 x 2.4m) sheet goods.

Squaring the Wall. Check walls to make sure that they are square before you stand them in place and again after you raise them. Measure the diagonals; they will be equal if the wall is square. It also helps to lock up the position by fastening diagonal braces along the wall.

Know the Terms

A stick-framed wall consists of studs—vertical 2x4s or 2x6s—between a single horizontal soleplate and a double top plate (top and cap plates). Windows and doors have framed headers and extra jack and cripple studs to account for the missing full studs. Doubled framing for doors (the full stud and jack stud) makes sense because doors can be heavy. Doubled studs around windows may not be necessary in a shed.

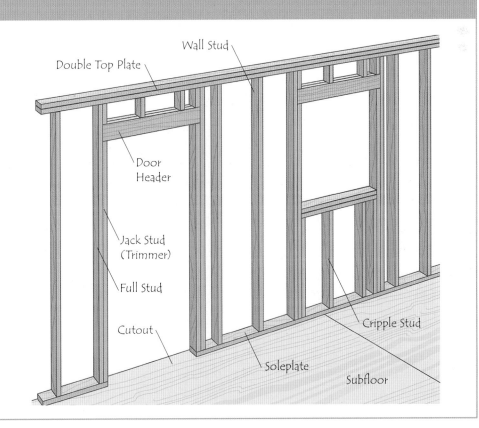

Double Top Plate · Wall Stud · Door Header · Jack Stud (Trimmer) · Full Stud · Cutout · Soleplate · Cripple Stud · Subfloor

Raising a Wall

Before erecting any wall, snap a chalk line along the sub-floor or slab to establish a reference guide for positioning the inside edge of the wall soleplate. On a subfloor, also fasten a few 2×4 cleats to the outside of the header joist to keep the wall from slipping off the deck as you raise it. With as many helpers as you need, slide the wall into position so that when you raise it, it will stand close to the guideline. Erect the wall, and align it to the chalk line.

Using a 4-ft. (1.2m) spirit level, get the wall as close to plumb as possible. You'll fine-tune it for plumb when you install the adjacent wall. Run braces from selected studs to cleats nailed into the subfloor or on corners. (See "Wall Bracing," opposite.) When the wall is plumb, have a helper nail the braces to the cleats. If you don't have a helper, nail an angled brace to a cleat on the deck and clamp it to the wall. Check for plumb, adjusting and re-clamping as needed; then fasten the brace to hold the wall plumb. On long walls of 20 ft. (6.1m) or so, brace each corner and several interior studs. With the bottom sole-plate properly positioned, nail into the header, rim, and/or floor joists using 16d nails.

Laying Out a Wall

To lay out the soleplate and top plate of a wall, first cut a pair of straight 2x4s or 2x6s to length. Tack the soleplate to the subfloor, and set the top plate flush against it. Make your first mark ¾ in. (1.9cm) short of your on-center spacing (15¼ or 23¼ in. [38.7 or 59.1cm]), and make an X past that mark. This will place the center of the first stud 16 or 24 in. (40.6 or 61cm) from the corner. Measure down the length of the plates, and mark with an X where each of the common, full-length studs will fall—every 16 or 24 in. (40.6 or 61cm). You can also measure and mark the locations of cripple studs (C) and jack studs (O).

Use a combination square to mark square lines for studs. (One wall plate is shown to emphasize the technique; you will mark two at the same time, as shown in the illustration below.)

Once you have drawn the square line, mark the location of a stud with an X. You can also use the tongue of a framing square, which is 1½ in. (3.8cm) wide, to mark the full width.

Lay out corners and rough openings on your plate, below. Mark all full-length studs with an X, jack studs (or trimmers) with an O, blocking with a B, and cripples with a C.

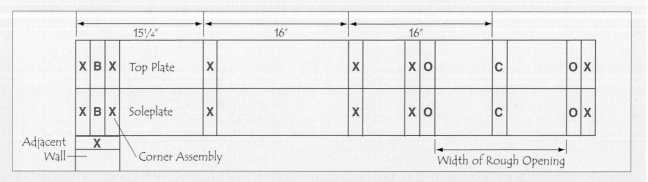

				15¼"		16"		16"					
X	B	X	Top Plate	X			X	X	O		C		O	X
X	B	X	Soleplate	X			X	X	O		C		O	X

Adjacent Wall — X — Corner Assembly ——— Width of Rough Opening

To check for plumb, hold a 2-ft. (61cm) or longer level against a straight 2×4. Pay particular attention to corners. If the wall is leaning in or out, release any braces and adjust the wall. Apply force to the braces to push the wall out. To bring the wall in, attach a flat, or spring, brace between two cleats (one attached to the wall and one to the floor) and use a two-by as a wedge board to bow the brace and force the wall inward (far right). You can also apply braces staked outside the building. Fasten the braces to hold the wall in its proper position.

Top Plates. When adjoining walls are in their proper positions, you can add the second top plate (also called cap plate) and tie the walls together. The cap plate on one wall overlaps the top plate of the adjoining wall. Where partitions join exterior bearing walls, the cap plate of the partition should lap onto the top plate of the exterior wall. Secure the laps using at least two 16d nails.

Wall Bracing

Site-built wall braces come in handy during construction. Use them to hold a wall in place or to force an uneven wall into plumb.

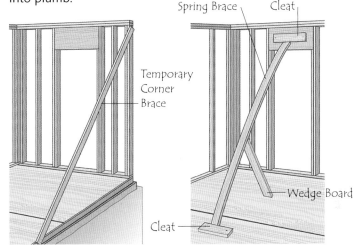

To brace a corner, lock it in place using a 2x4. Leave the brace in place until the whole wall is finished. Nail it outside the frame so that it is out of the way.

Apply a spring brace to fix a bow in a stud wall. Nail a flat brace (at least 8 ft. [2.4m] long) to cleats on the floor and wall, and force the wall into plumb.

Stud Configurations at Corners

A two-stud corner is usually fine for a small shed with unfinished interior walls. If you plan on finishing the interior of your shed, you will need to provide a nailer at the corners for the finish material, such as plywood or paneling. A stud-and-block corner uses the most material, but it is the most rigid. A three-stud corner saves time and a little material because you don't need the blocking.

Two-Stud Corner

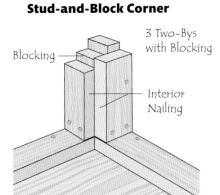

2 Two-Bys (No Blocking)

Interior Nailing on One Edge Only

Stud-and-Block Corner

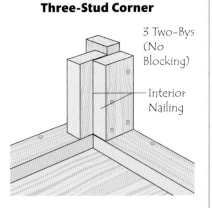

3 Two-Bys with Blocking

Blocking

Interior Nailing

Three-Stud Corner

3 Two-Bys (No Blocking)

Interior Nailing

Windows and Doors

Stud placement is crucial for rough openings in walls where you insert doors and windows. The rough-opening size is listed in window and door spec sheets, including about ½ to ¾ in. (1.3 to 1.9cm) of shimming space around the unit. This allows you to plumb and level the window or door even when the wall frame is less than perfect. Each side of the opening has a full-height stud. Inside that stud is a shorter jack stud. The distance from jack to jack, allowing for shimming space, is the rough opening.

On windows, a short jack stud helps to support the sill. You add a jack stud that runs from the sill up to the header. On most sheds, there is no need to install a beefy header above the window as shown below. Often you can frame the opening using doubled 2×4s, but check with the local building inspector for requirements in your area.

On doors, you need two long jack studs, which run from the floor to the header, with cripple studs above. You can rest the jacks on the soleplate or run them down to the plywood deck.

Building Rough Openings

1. Nail a jack stud into the soleplate and adjacent full-height stud at each side of the rough opening. In 2x4 walls, use two l0d nails every 12 in. (30.5cm) or so for stability.

2. Continue additional jack-stud sections along both sides of the window opening. These framing members will help to support the weight carried by the header across the opening. With small sheds you may be able to omit the jack studs.

3. Make up a header with two 2x6s or 2x4s, sandwiching ½-in. (13mm) plywood to beef up the header to the wall thickness. Wider openings may require larger headers.

4. Add cripple studs above the header and below the sill to maintain the on-center wall layout. You need the cripples for support and as nailers under surface materials.

Framing Gable Roofs

If you are building a simple single-plane slanted roof like that on the goat shed shown on pages 194-205, you can skip this section altogether. If you are building a shed with a gable roof, however, you will need a review of the basic geometry required for planning the project and buying materials.

Calculating rafter length and the angles at the ridge and rafter tail is more complex than framing walls. But the job is doable using only basic math if you break down the task into smaller, simpler steps. On gable roofs, you can either run the rafters individually to a central ridgeboard or assemble gable-style trusses on the ground and lift them into position. Some lumberyards carry prefabricated trusses for several basic roof spans and pitches.

Because all common rafters in a gable roof are the same, you can mark and cut one, test the fit, and use it as a template. Most rafters get three cuts: a plumb cut at the ridge where the rafter rests against the ridgeboard; a plumb cut at the tail, which makes the shape of the bottom end; and a bird's-mouth cut where the rafter sits on the plate of the outside wall. Sometimes you need a fourth cut, a horizontal cut at the tail, which is often used with an overhang and soffit.

Determining Gable Height. If you want to know what the final height of your roof will be, you can determine it if you know your shed's slope and span. Begin by dividing the span in half to get the total run. (See the illustration below.) Let's say that your shed has a span of 14 ft. (4.3m). The total run is 7 ft. (2.1m). Now multiply the unit rise (the rise per foot of run) by the number of feet in the total run. For example, an 8-in-12 roof has a rise of 8 in. (20.3cm) for every foot (30.5cm) of run. An 8-in-12 roof with a total run of 7 ft. (2.1m) has a total rise of 56 in. (8 × 7), or 4 ft. 8 in. If you increase the run, the slope doesn't change, but the total rise increases. For example, if you have a 10-ft. (3m) run with an 8-in-12 roof, the total rise is 80 in. (8 × 10), or 6 ft. 8 in. (203.2cm).

When you determine rise, you use a measurement along a line from the top plate's upper outside edge to the ridge's centerline. The point at which the rafter measuring line and the ridgeboard centerline intersect is known as the theoretical ridgeboard height. The rise is the distance from the plate to the theoretical ridgeboard height.

Gable Roof Framing

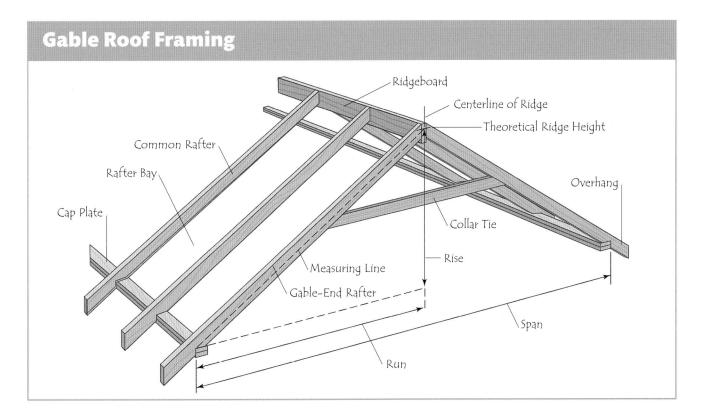

Roof Terminology

Rise is the height of the roof at its theoretical ridge height measured from the top plate of the end wall below the ridge.

Span is the horizontal distance from wall to wall, measured outside. A roof's span does not include the overhang at the eaves.

Run is the horizontal distance from one wall to a point under the middle of the ridge, or typically half the span.

Slope is expressed as the rafter's vertical rise in inches per 12 in. (30.5cm) of horizontal run. A roof that rises 4 in. (10.2cm) for every 12 horizontal inches is expressed as 4:12 or 4 in 12. On most building plans, you'll notice a right triangle off to the side—for example, 4 in 12 with a 12 at the top of the triangle on one leg of the right angle and a 4 on the other leg. The hypotenuse of the triangle shows you the angle of slope. The higher the number of inches in unit rise, the steeper the roof. A 12-in-12 roof, common in Cape Cod-style roofs, rises a foot in elevation for every foot of run (a 45-degree angle).

Roof Trusses

Rather than cutting your own roof rafters, you can use roof trusses. For sheds, trusses are usually 2x4s held together by metal or wooden gussets. Order them from a truss manufacturer by specifying the length of the bottom chord, which includes the span of the shed, the thickness of the walls, and the soffits or overhangs if any. The manufacturer will tell you the number you will need for your shed. You will probably need a helper to lift and install the trusses.

Roof Harmony

If you'd like the roofline of your shed to match your house, here is a sure-fire method to capture the rise and run. Hold a combination square and a ruler as shown. Use the bubble in the combination square—or the edges of horizontal siding if your home has it—to level the ruler. Slide the combination square until it lines up with the 12-in. (30.5cm) mark on the ruler. Now slide the combination square's ruler up until it intersects the roofline. You now have the amount of rise per 12 in. (30.5cm) of run—all without getting up on a ladder.

Estimating Rafter Size

Before you order rafter lumber, you need to know what size boards to get. That means determining how long a rafter you'll need to cover the shed, and then adding on the amount of overhang (if any) you want at the eaves.

Let's say you plan a slope of 8 in 12, which means the roof rises 8 in. (20.3cm) for every 12 in. (30.5cm) of run. The width of the shed is 12 ft. (3.7m). The run is one-half the building's width, in this case, 6 ft. (1.8m). Once you know the slope and run, you know that the roof will rise 4 ft. (8 × 6 = 48 in., or 4 ft.). You are now ready to determine the rafter length. If you think of half the roof as a right triangle, you already know the base (6) and altitude (4). You need to figure out the hypotenuse of this right triangle, which represents the rafter length. Using the Pythagorean theorem:

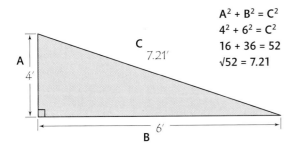

$$A^2 + B^2 = C^2$$
$$4^2 + 6^2 = C^2$$
$$16 + 36 = 52$$
$$\sqrt{52} = 7.21$$

Don't be spooked by the square root. Even the simplest of pocket calculators includes a square-root function. The square root of 52 ft. (15.8m) is 7.21 ft., which equals 7 ft. 2½ in. (2.2m). If your rise and/or run are not in whole feet but in feet and inches, then convert the whole figure to inches; do the math; and convert it back to feet. Use decimals of a foot rather than inches when you divide the resulting number of inches by 12 on a calculator to arrive at feet.

Next, you'll need to determine the amount of overhang you want. The overhang is the level distance from the edge of the building, but the actual rafter length is longer because of its slope. You can again use the Pythagorean theorem to figure out the dimension to add to the rafter length for the overhang. If you want an 18-in. (45.7cm) overhang on the same 8-in-12 roof, for example, you must envision the overhang area as a miniature roof. The run is 18 in. (45.7cm) (the horizontal dimension of the overhang) and the rise is 12 in. (30.5cm) (8 × 1.5 = 12). Therefore,

$$12^2 + 18^2 = C^2$$
144 in. + 324 in. = 468 in.

The square root of 468 in. (1,188.7cm) is 21.63 in., which is 1.80 ft., or 1 ft. 9⅝ in. (54.9cm). Add this to the rafter length already determined:

7 ft. 2½ in. + 1 ft. 9⅝ = 9 ft. ⅛ in. Because lumber is sold in 2-ft. (61cm) increments, you'll need to order 10-ft. (3m) (or 20-ft. [6.1m]) pieces of lumber for your rafters.

Rafter Cuts

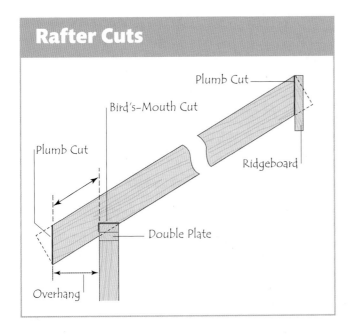

Gable Rafter Layout

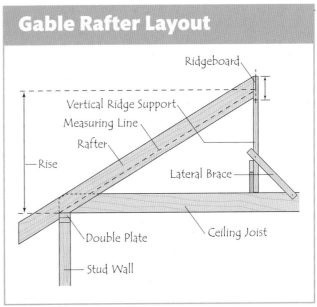

Saltbox Shed

Building the Shed

This large shed offers plenty of space for feed, straw, and tool storage—with room left over for a potting bench and a small shop. The double doors ease access and boost ventilation in the summer; windows offer plenty of natural light. And it is a permanent structure, markedly different from a metal prefab in appearance, spaciousness, and construction. It will last as long as your house does.

This shed is a major project, of a magnitude that your personal safety could be at risk as you raise walls and complete the roof. Enlist experienced help when you can. Another tip: consider assembling the frame with 3-in. (76mm) deck screws. That way, you can back out of any mistakes.

Reality Check

Once you have roughed out a site for the new shed, double-check that the location meets any setback requirements that apply. If you are excavating, make sure there are no buried water pipes or electrical lines in the area.

Saltbox Shed, Perspective View

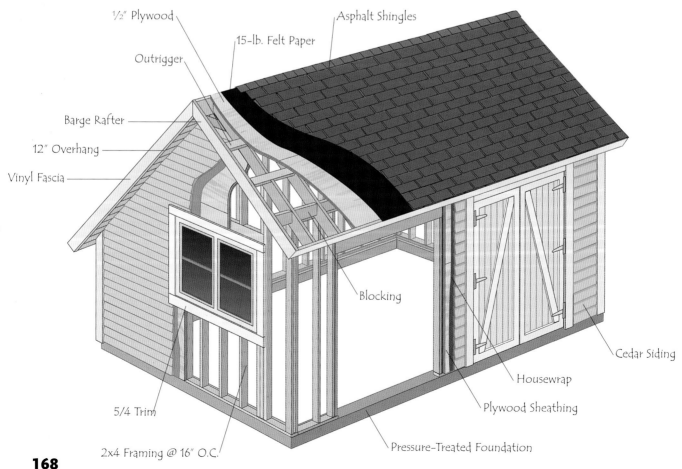

½" Plywood

15-lb. Felt Paper

Asphalt Shingles

Outrigger

Barge Rafter

12" Overhang

Vinyl Fascia

Blocking

Cedar Siding

Housewrap

5/4 Trim

Plywood Sheathing

2x4 Framing @ 16" O.C.

Pressure-Treated Foundation

Design Basics. The shed measures 12 × 16 ft. (3.7 x 4.9m), with two pairs of wide, hinged doors and windows in both end walls. The walls are framed using 2×4s, with ½-in. (12mm) plywood or OSB sheathing. The siding is beveled western red cedar, but fiber cement siding (Hardiplank) or even the plywood sheathing alone would serve as well. The painted wood trim is ¾ pine.

The shed is built on a foundation of 6×6 pressure-treated timbers set on a gravel footing. The floor is compacted screenings, or fine gravel. You can wet the gravel down and compact it to form a smooth, hard surface. This avoids the expense of a concrete slab and footings.

Framing the Roof. Setting up your ridgeboard and attaching rafters is the trickiest and most hazardous phase of building this shed. You'll do better work and do it more safely if you rent a scaffold. Give yourself a solid working platform by placing it in the center of the shed before you start installing rafters.

Materials

Foundation
8 16-ft. (4.9m) 6x6s
8 12-ft. (3.7m) 6x6s
84 10-in. (25.4cm) galvanized spikes
20 24-in. (61cm) rebar spikes
Gravel
Screenings or fine gravel

Framing and Sheathing
28 4 x 8-ft. (1.2 x 2.4m) sheets of
     ½-in. (1.3cm) CDX plywood
2 10-ft. (3m) 2x10s
19 12-ft. (3.7m) 2x8s
19 8-ft. (2.4m) 2x8s
3 12-ft. (3.7m) 2x6s
7 16-ft. (4.9m) 2x4s
5 14-ft. (4.3m) 2x4s
12 12-ft. (3.7m) 2x4s
28 10-ft. (3m) 2x4s
11 8-ft. (2.4m) 2x4s
4 10-ft. (3m) No. 2 pine 1x8s
8d cement-coated nails
12d cement-coated nails
16d cement-coated nails

Roofing
9 10-ft. (3m)-long sections of
     aluminum drip edge
3 rolls of 15-lb. builder's felt

13 bundles of shingles (3 bundles
     per square)
⅞-in. (2.2cm) roofing nails
⅝-in. (1.6cm) staples
2 tubes of asphalt cement

Windows
4 34 x 36-in. (86.4 x 91.4cm)
     double-hung windows

Trim Boards
15 12-ft. (3.7m) ¾x6s
6 14-ft. (4.3m) ¾x6s
6 12-ft. (3.7m) 1x6s
6 10-ft. (3m) 1x8s
12d finish nails
3 tubes of latex caulk
1 gal. of latex exterior primer
1 gal. of latex exterior paint

Soffits and Fascia
13 12-ft. (3.7m)-long sections of
     vinyl soffit
8 12-ft. (3.7m)-long sections of
     vinyl fascia
8 vinyl J-channels
Aluminum trim nails

Siding
1,060 lin. ft. of ½ x 8-in.
     (323.1m of 1.3 c 20.3cm)
     western red cedar beveled
     siding
2-in. (5.1cm) aluminum siding nails
1¼-in. stainless-steel screws
1 gal. of latex semitransparent
     exterior stain

Doors
4 4 x 8-ft. (1.2 x 2.4m) sheets of
     ¾-in. (1.9cm) exterior plywood
13 8-ft. (2.4m) No. 2 pine 1x6s
1 8-ft. (2.4m) No. 2 pine 1x8
2 tubes of construction
     adhesive
6d finishing nails
12 T-hinges
48 3-in. (76mm)#12 screws
48 ⁵⁄₁₆ x 2-in. (8 x 51mm) hex bolts
     w/washers, split washers,
     and nuts
2 clasps and staples
     w/mounting screws
2 sliding bolts w/mounting
     screws

Saltbox Shed

Building the Foundation

The saltbox shed is built on a perimeter foundation of 6x6 pressure-treated timbers spiked together with 10-inch galvanized nails. It is important to promote drainage away from the foundation and, in cold climates, to extend it below the frost line. Hire an excavator to dig the foundation trenches and strip the sod from the interior area using a backhoe.

Spread gravel in the trenches; then lay the first course of timbers. The number of courses needed depends on how deep your trenches are. You want the top surface of the foundation just a few inches above grade. Take time to level and square the first course accurately. The adverse effects of a pitched, out-of-square base will compound themselves as you frame.

1

Lay out the foundation corners by using two measuring tapes and the 3-4-5 method. When the 16- and 12-ft. (4.9 and 3.7m) sides are laid out at right angles, the diagonal measure will be exactly 20 ft. (6.1m).

2

Mark the outside corners of the foundation trenches on the ground using spray paint or flour. Nothing more is needed for an excavation job this simple. Dig the trenches.

3

Shovel in a layer of gravel about 6 in. (15.2cm) deep to promote drainage. Level the gravel's surface as much as possible before setting the first-course timbers in place.

Drive rebar pins through the first-course timbers to tie them to the ground. Bore a pilot hole completely through the timber; then hammer the 24-in. (61cm)-long pin in place using a hand sledgehammer.

Build up the foundation, with the timbers alternating at the corners log-cabin style. Position all four timbers for a course, and measure the diagonals to ensure that the course is square. Drill a pilot hole; then spike the new course to the one below using l0-in. (25.4cm) galvanized nails.

Backfill the trenches as soon as you can. Shovel a layer of soil into the gap between the trench wall and the foundation; then tamp it. Backfill more, and tamp again. The soil will settle, so backfill a few inches above grade.

Spread the screenings or fine gravel; then level and tamp them. Have the material dumped directly into the foundation, if possible. As you spread and level the screenings or gravel, tamp it methodically, either by hand, as shown, or using a rented power tamper.

Continued

Saltbox Shed (cont'd)

Framing the Back Wall

In general, you frame a wall by cutting the parts, laying them out on a flat surface, and nailing them together. Nail the sheathing to the frame while it is lying flat because it's easier that way and lets you square the wall before erecting it.

Choose long, straight 2x4s for the plates. (See at right for how to lay out the plates.) Cut the studs and trimmers to length, and make up the corner posts and any headers needed. Lay the frame parts on the shed floor; line up the studs and plates; and drive 12d nails through the plates into each stud.

Measure the frame, and cut pieces of ½-inch CDX plywood to size. (You will cut out windows and doors after fastening the sheathing.) Extend the plywood 1 inch beyond the bottom plate. Size the pieces so that joints will occur over a stud. Leave a ⅛-in. (3.2mm) gap between pieces.

Next, set up and brace the wall with a helper or two. The wall is heavy, and once it is upright, you will need at least one person to hold it up while the other gets the bottom aligned properly on the foundation and sets up the bracing.

Once you have braced the wall, drive 16d nails through the soleplate into the foundation.

Back Wall Frame

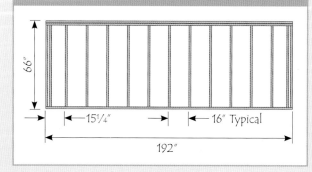

66"

15¼" 16" Typical

192"

8

Lay out the top plate and soleplate for the back wall together. Measure and mark the stud locations, and mark an "X" on the side of the line where the stud will be attached.

9

Cut the studs by piling identical 2x4s with their ends flush. Mark a cut line on the top layer of studs. As you cut through the top studs, you'll kerf those underneath, marking where you need to cut them.

10

Line up the plates and studs, and drive 12d nails through the plate into the end of each stud. Use foot pressure to align the top surfaces as you drive the nails.

11

Take the diagonal measurements to ensure that the frame is square. You can shift the alignment of the assembly by rapping a corner with a sledgehammer. Keep at it until both diagonal measurements are exactly the same.

12

Attach the plywood sheathing. Mark the location of the studs for easy nailing. The sheathing prevents the wall from shifting out of square as you lift it into position.

13

Set up the frame; plumb it; and add braces immediately. The job requires at least two: one to hold the wall and check the level, the other to nail the brace to the foundation.

Continued

Saltbox Shed (cont'd)

Framing the Front Wall

14

Mark the top plate and soleplate together. On each, mark only the locations of studs that actually abut it. Code the location according to the kind of stud.

15

Build up a header by face-nailing two 2x8s together, with a ½-in. (12mm) plywood spacer between them. The plywood doesn't have to be a continuous piece; this is a good opportunity to use up leftover scraps.

16

Face-nail the trimmer, or jack, studs to full-length studs, and make up corner posts by face-nailing studs together. You can economize by sandwiching short pieces of 2x4 between the studs to form the corner posts.

17

Assemble the walls by laying the frame parts on the shed floor for assembly. One by one, line up the parts on the appropriate layout lines, and nail through the plates into the ends of the studs.

18

Sheathe the wall frame after squaring it. Apply full sheets of plywood, covering up the top portion of the doorways. After you have sheathed the wall, snap cut lines and cut out the doorways.

19

Raise the wall into position. It is heavy; you'll need a helper or two to tip it up. Once you have tipped up the wall, it takes little effort to hold the assembly upright.

20

Install temporary bracing to hold the wall erect. Nail 2x4s to the corner posts and to scraps scabbed to the foundation, as shown. Make sure the wall is plumb.

Continued

Front Wall Frame

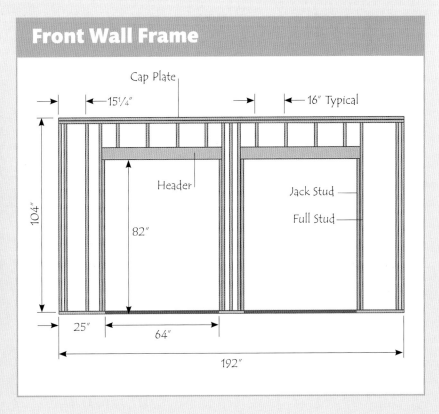

Cap Plate

15¼"

16" Typical

Header

Jack Stud

Full Stud

104"

82"

25"

64"

192"

Saltbox Shed (cont'd)

Framing the Roof

The biggest challenge in building the saltbox shed is framing the roof. Due to the gable-end overhang, the 18-foot ridgeboard is longer than readily available standard 2x8s, which means that you'll have to join two 9-foot pieces end to end—and make the connection 12 feet above the sill. Also, because of the saltbox style, the front rafters are shorter than the back ones, so you must cut two rafter lengths.

Deal with the ridge first. Get it up; then you can cut and fit front and back rafters. Use the first two rafters as patterns for cutting the remaining ones.

Unless you have several helpers, it's easiest to splice the ridgeboard after you hoist the two sections into place atop temporary supports. Lay out the rafter locations while the ridgeboard sections are on the sawhorses. Select straight boards, and prepare two splice boards, which you'll scab across the end-to-end butt joint once the sections are aloft. The splice must fall between rafters.

Setting up temporary supports for the ridgeboard requires ingenuity. Because you haven't framed the sidewalls yet, you have to stand a long stud on the foundation at each wall and attach braces to hold them erect. You must erect and brace a similar stud midway between the end supports. The supports must not interfere with the rafters or the splice.

Stretch a level line across the support studs, and scab a 2x4 to each stud at the line. Rest the ridge on these 2x4s, and clamp it to the support studs. After clamping the ridge, attach the splice boards.

There are a couple of methods for laying out rafters, but the best for this project is the step-off, which uses a framing square. Follow the drawings at left.

You must notch two front and two back rafters for outriggers to which the barge rafters are attached. The barge rafters form the gable-end roof overhangs. As you install the rafters, place the notched ones flush with the ends of the walls. Set the outriggers into the notches, and nail through the adjacent rafters into their ends. You will nail the barge rafters to the ridge and the outriggers.

Complete the roof framing by nailing blocking between the rafters, flush with the outside of the front and back walls. Face-nail through the rafter for one side of the blocking, and toenail the other side. Cut and install the collar ties. Last of all, cut and nail fascia boards spanning the overhanging ends of the rafters at the front and back.

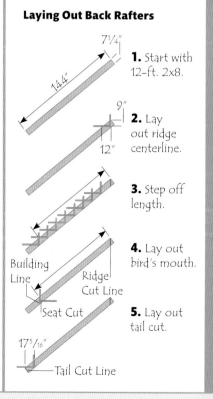

Laying Out Front Rafters

1. Start with 8-ft. 2x8.

7¼"
96"

2. Lay out ridge centerline.

9"
12"

3. Step off length.

4. Lay out bird's mouth.

3½"

5. Lay out tail cut.

4"

Laying Out Back Rafters

1. Start with 12-ft. 2x8.

7¼"
14·4"

2. Lay out ridge centerline.

9"
12"

3. Step off length.

4. Lay out bird's mouth.

Building Line
Ridge Cut Line
Seat Cut

5. Lay out tail cut.

17⁵⁄₁₆"
Tail Cut Line

21

Mark the ridge centerline; offset it by ¾ in. (1.9cm) (half the ridge thickness); and mark the ridge cut line. Then, starting from the ridge centerline, step off along the rafter in 12-in. (30.5cm) increments until you reach the "building line"—the outer edge of the framing.

22

Lay out the seat cut by aligning the framing square's blade (the long end of the square) on the building line and the 4-in. (10.2cm) mark of the tongue (the short end) with the rafter edge. Scribe along the tongue.

23

Make a front and back rafter, and try them in place. When you've got them cut right, use each as a template for laying out the others. Align the rafter on a 2x8, and trace the ridge cut and the bird's mouth—the notch that sits on the framing.

24

Mark the rafter locations on the ridge. Measure from the scarf-joint bevel on each of the two 2x8 ridge sections. Mark both sides of the ridge.

Continued

Saltbox Shed (cont'd)

Framing the Roof

Raising the Ridge

91³/₁₆″   43¹³/₁₆″

9¹/₄″

Cleat nailed to 2x4 supports ridge.

136¹¹/₁₆″

2x4 Temporary Support for Ridge

90³/₁₆″   45⁵/₁₆″

137″

25

Erect temporary supports for the ridge. Use chalk-line string to align the "seats" for the ridge atop the supports. Clamp the two ridge sections onto the supports.

26

Install the rafters by face-nailing the first rafter that abuts the ridge at each location. Drive three 16d nails through the ridge into the end of the rafter. Add the second rafter at each location by driving a nail through the top edge into the ridge (inset) and toenailing each side.

27

Toenail each rafter to the top plate using 12d nails. Drive nails into both faces of the rafter, angling each down into the plate.

28

Close in the spaces between rafters with blocking. Rip a 2×8 so that it extends from the wall plate to the top of the rafter; then crosscut it to fit between the rafters. Face-nail through one rafter into the blocking, and toenail through the other.

29

Install each collar tie by first resting it on the front top plate. Level and face-nail it to the back rafter. Then go back to the front end and face-nail the tie to the rafter and toenail it to the top plate.

30

Add the outriggers by clamping front and back rafters face to face; then cut the outrigger notches in both at the same time. Remove the bulk of the waste using your circular saw. Then chisel the bottom of the notches flat.

31

Fit the outriggers into the notches, and pull their ends tight to the adjacent rafter. Drive two 16d nails through the outrigger into the rafter. Face-nail through the adjacent rafter into the end of the outrigger.

Continued

Saltbox Shed (cont'd)

Framing the End Walls

32 Temporarily brace the front and back walls during construction. Nail a scrap to the wall; then butt a brace underneath it. Nail the brace to a stake in the ground.

33 Mark each end-wall stud by standing it at the penciled location on the soleplate and marking along the underside of the rafter on its edge. This will be the bottom of the notch you need to make. Use a level to check that the stud is plumb as you mark it.

34 Cut the 1½-in.-wide (3.8cm) notch using your circular saw. Begin by making the angled shoulder cut across the stud's edge, and then rip in from the end to that cut.

35 Toenailing a stud to the soleplate is easy if you brace the stud against a temporary block. Then the hammer blows won't move the stud off of its mark.

36

Plumb the stud; then drive a couple of 12d nails through the stud into the rafter.

37

Add cripple studs, which extend from the soleplate to the rough sill beneath the windows. After you have set the sill, stand the cripples one by one on the soleplate layout marks, and face-nail through the sill into their ends. With the tops thus secured, it is easier to toenail the bottom ends to the plate.

38

Make the header out of two 2×8s and some plywood scraps. Install trimmers that extend up from the rough sill to hold the header. Set the header in place and, using 16d nails, face-nail through the full-length stud into it.

39

Add cripple studs between the header and the rafters. Mark and notch these, and stand them in place. Face-nail them to the rafter; toenail them to the header.

Continued

Saltbox Shed (cont'd)

Sheathing the End Walls

40

Temporarily nail a ledger to the foundation to support the ½-in. (12mm) plywood sheets at the correct height. Install full sheets first. Set a sheet on the ledger; align it left and right; then nail it to the studs.

41

Sheathe the triangular gable end. With the first three sheets in place, take measurements from the shed to lay out the triangle. Measure it to fit against the underside of the outriggers. Mark a notch for the ridge; cut the piece; and install it.

42

Cut the rough openings for the windows and doors by first drilling a hole through the sheathing from inside the shed at each corner of the opening. Draw or snap lines from hole to hole; then cut along the lines to remove the sheathing.

Sheathing the Roof

43

Install the barge rafter, which sits beyond the sidewall, attached to the ends of the outriggers. Use ¾ pine for appearance's sake (shown), or pick out the most knot-free and straight 2x8 you can find.

44

Nail the fascia to the ends of the rafters. Join the fascia boards in a scarf joint, and locate the joint over the end of a rafter so that you can fasten it securely.

45

Begin sheathing the roof along the eaves. After nailing the first sheets down, nail a 2x4 to the rafters to provide footing. Add 2x4s as you install the second row of plywood sheets, and so on up the roof. Be sure that the ends of each sheet are halfway over a rafter so that you can securely fasten the sheets using 8d nails. Nail every 6 in. (15.2cm) along the end seams, every 12 in. (30.5cm) in the field.

Continued

Saltbox Shed (cont'd)

Shingling the Roof

Straightforward roofing like this requires only a measuring tape, chalk-line box, hammer, utility knife with extra blades, speed square, stapler for the roofing felt, and metal shears to cut the drip edge. A rented pneumatic roofing nail gun speeds up the work, but on a shed this size, the rental cost and the time spent picking it up and dropping it off are barely worth the labor saved.

From both safety and efficiency standpoints, scaffolding is essential because you want to be able to work along the eaves. Roofing jacks (Steps 51 and 52) are a worthwhile investment.

Roll roofing is a tempting option for a shed, but you should avoid it. Though slightly less costly than shingles, it won't last as long and is surprisingly hard to install neatly.

46

Staple 15-lb. builder's felt to the roof. Cut the strips to span the full width of the roof. Align the first strip with the bottom edge of the roof, and staple it down. Lap the second course over the first by 4 to 6 in. (10.2 to 15.2cm). Don't over-staple; fasten the felt just enough to hold it in place until you shingle.

47

Install drip edge to protect the edges of the sheathing. Along the lower edge of the roof, the roofing felt overlaps the drip edge. Along the gable edges, install the drip edge overlapping the felt.

48

Begin shingling the roof using a starter course. Cut the tabs off shingles, and position and fasten them along the roof edge.

49

Lay the first full course of shingles so that the tabs completely overlay the starter course. Overlap any seams. Follow the nailing pattern specified by the manufacturer. Typically, you use four nails per shingle, positioning them above each notch between tabs.

50

Begin the next several courses. You need to stagger the notches between tabs from course to course, so cut down the first shingle in five of every six courses. The second course begins with a 2½-tab shingle, the third with a 2-tab shingle, and so on.

51

Add roofing jacks to work safely as you progress up the roof. For each jack, drive a 16d nail through the body of a shingle and into a rafter. Leave the head protruding so that you can catch the jack on the nail. Two jacks are sufficient to support an 8-ft. (2.4m) 2x6.

52

Remove the jacks when the roofing is complete by lifting the tab and unhooking the jack. Seat the nail, and cover the head with a dab of roofing asphalt. Push the tab down.

Continued

Saltbox Shed (cont'd)

Installing the Windows

53

Wrap the shed in builder's felt or house wrap, extending the strips around the corners and across the rough openings for the windows. Each strip overlaps the one below it. Strips lap end to end, too. Staple the felt to the plywood. At the rough openings, slice through the felt using a utility knife.

54

Fold flaps of the felt over the edges of the rough window openings, and staple them to the studs, sill, and header.

55

Install the windows in the rough openings. Stand a unit on the sill, and align it laterally. Then tip it into the opening. Make sure that the nailing flange seats against the shed walls all around. Have a helper inside the shed plumb the unit using shims.

56

Fasten the windows by driving nails through the nailing flange into the framing of the rough opening. Most windows have punched holes in the flange to indicate where to locate the nails.

Trimming the Soffits and Fascia

57

Enclose the eaves using vinyl soffit panels (shown) or strips of ¼-in. (6mm) exterior plywood. Under the front and rear overhangs, the panels run across the rafters. Nail them to the bottom edges of the rafters. At the gable ends, fasten two-by nailers to the sidewalls parallel with the barge rafters.

58

Cut and mount short pieces of soffit to the gable-end eaves. Cut the vinyl using a saber or circular saw. Hook the bottom of the soffit into the interlock bead on the previously installed piece. Slide the piece along the bead to the wall.

59

Attach the soffit. Pull the piece tight, ensuring that it is hooked to its neighbor; then drive aluminum or galvanized roofing nails through the nailing slots. Driving one nail into the nailer strip and one into the rafter will hold it.

60

Cover the barge rafters and fascia boards using vinyl with a low-maintenance finish. The lip on the vinyl fascia strip overlaps the soffit ends, and the top edge slips under the roof drip edge.

Continued

Saltbox Shed (cont'd)

Trimming the Windows, Doors, and Corners

The trim applied to the saltbox shed is a significant part of its traditional design. You will frame the windows and doors using wide boards and apply matching boards at the corners. To produce the desired appearance, use ¾ stock for this trim. Typically, ¾ pine is 1 to 1³⁄₁₆ inches thick and stocked in standard nominal widths from 4 to 12 in. (10.2 x 30.5cm).. You may be able to order pre-primed ¾ trim boards.

You will apply the trim after wrapping the shed with felt paper or housewrap and trimming out the soffits. Cut and fit the boards tightly around the windows and doorjambs. You will probably get the best fit around the windows if you cut rabbets in the trim to accommodate the nailing flange of the window. As you cut each part, tack it in place to test-fit it.

Before finally nailing the trim, pull it down and prime the boards if not pre-primed—front, back, ends, and edges. When the primer is dry, nail up the trim. Chalk the seam between the trim and the windows. Apply at least one finish coat to all exposed surfaces. The shed is ready for siding.

The Easy Way to Prime

Prime raw-lumber trim before nailing it in place. It is best to apply a coat of primer on both front and back and the edges, including the end grain. The primer acts as a sealer that helps prevent the wood from absorbing moisture.

61

Install the corner trim. Miter the board that butts the gable-end soffit, and bevel the adjoining board at the same angle. Nail the two pieces together.

62

Fasten the corner trim using 12d galvanized finishing nails, making sure that the assembly is square against the shed walls and tight against the soffits. Check that the nails penetrate the corner posts of the shed.

63

Cut out the soleplate before casing the doorways. Use a handsaw, cutting flush with the studs on each side of the doorway.

64

Cut and fit the window trim; then prime it. If necessary, mill a rabbet into the back of each trim strip to accommodate the nailing flange so that the board lies flat against the sheathing. You may need to rabbet the face to fit the board into the window's siding channel.

65

Set the header trim on top of the windows. Align it laterally, and seat it tightly against the windows. Fit the vertical trim on each side of the windows. Seat it tightly against the window and against the header.

66

Nail the trim to the shed using 12d galvanized finishing nails. Use a nail set to countersink the heads. Finish the installation by filling the nailholes with caulk before applying a finish coat of paint.

Continued

Saltbox Shed (cont'd)

Siding the Shed

Beveled wood siding is especially attractive and reasonably durable, although it is expensive. For a backyard-homestead shed, fiber-cement or hardboard lap siding will look as good. Sheet goods—exterior plywood, fiber-cement panels, and pressed hardboard or OSB (oriented-strand board)—with vertical 1x2 battens are another natural choice, as is T1-11 plywood siding. Or consider cedar shingles. Vinyl siding will be savaged by your animals, so you should avoid it.

Western red cedar siding, which weathers extremely well, is shown in these steps. The siding is ½ in. (1.3cm) (at the thicker edge) and 8 inches wide. Typically, it is delivered in random lengths, ranging in 1-foot increments from 4 ft. (1.2m) up to 16 ft. (4.9m).

Cedar splits easily, so exercise care in handling and cutting it. Drill a pilot hole for every fastener you drive.

You can paint or stain cedar, and you should apply stain or primer to the front and back before installing the siding. If you leave it natural, it will darken markedly as it weathers.

67

Begin siding the shed by nailing a starter strip along the bottom of the wall. Rip the strip to a height equal to the siding overlap. Use the thin top edge of damaged siding pieces for the starter strip, and align it flush with the bottom ends of the corner boards.

68

Install the first strip of siding over the starter strip. Cut it to fit snugly between the corner boards. You might choose to use two pieces to form a single course here and there as you advance up the wall, but it's best to use a single, continuous strip for the first course.

69

Avoid splintering while cutting cedar by scoring the cut line with a utility knife. This practically eliminates splinters, especially handy when cutting prefinished siding. A power miter saw makes fast crosscuts, but using a saber saw also gets the job done and is useful for cutting notches.

70

Notch the siding around windows and doors. A single siding strip to fit beneath and above windows and doors takes careful cutting. Layout must be accurate, and the strip can be difficult to maneuver into position without breaking. But the payoff is a better-looking job. The alternative is to use two strips, one on each side of the window, to make up the course.

71

Drill a pilot hole for every nail you drive through the siding. Cedar splits easily. A simple gauge aligns the bit in relation to the edge so that all of the nailheads line up. Snap chalk lines on the builder's felt to mark stud locations.

Nailing Siding

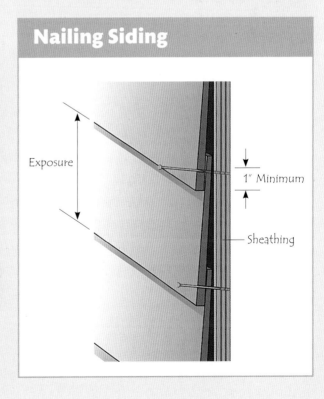

Exposure

1" Minimum

Sheathing

Continued

Saltbox Shed (cont'd)

Building and Hanging the Doors

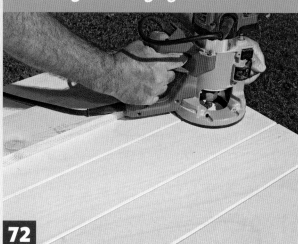

72

Build each door by attaching a wooden frame to a ¾-in. (19mm) plywood panel. For a decorative touch, use a router and V-grooving bit to cut grooves in the panel's face after cutting the plywood to size. Clamp a straightedge to the plywood, and guide the router along it.

73

Cut the door-frame members to size. Use 1x6 for the sides and top, 1x8 for the bottom. Set the pieces in place, and clamp them. Crosscut the 1x6 for the diagonal brace a couple of inches longer than needed, and rip it to 4½ in. (11.4cm). Position the roughly sized diagonal brace, and clamp it.

74

Mark the intersection of the brace and the horizontal frame piece. Set a sliding bevel to the angle as shown— the tool's body against the horizontal piece, its blade against the brace. Use this angle to scribe a cut line from the mark across the brace piece.

75

Unclamp the frame parts, one by one; apply construction adhesive to their backs; and then reposition all of them before permanently fastening them in place.

76

Clamp the parts to one another and to the plywood while the adhesive sets. A pipe clamp applied across the vertical frame members pinches them against the ends of the horizontal parts. Bar clamps and C-clamps squeeze the parts against the face of the plywood. While the parts are clamped, drive and countersink 3d galvanized finishing nails.

77

Apply shellac to all knots before priming and painting the doors. Use a dewaxed or pigmented shellac for this job. These sealers, which dry quickly, prevent knots from bleeding through the paint. Brush a couple of coats over the knots and pitch pockets only; then prime and paint the entire door after the shellac dries.

78

Place the doors, with T-hinges already mounted on them, into the doorways to mark the hinge locations on the doorjambs. A permanent stop nailed to the head jamb and a temporary one tacked to the sill help position the door. Set the bottom of the door on a shim, and slide it against the side jamb for marking and installation using the screws provided with the hinge.

Door Layout

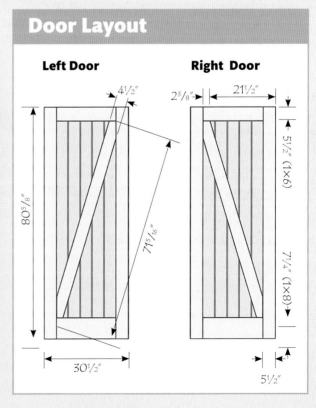

Left Door — Right Door

4½" 2³⁄₈" 21½"

80⁵⁄₈"

71⁵⁄₁₆"

30½"

5½" (1×6)

7¼" (1×8)

5½"

Goat Shed

Goats don't ask for much in the way of shelter. They are amazingly hardy creatures whose only Achilles' heel is dampness. They can take extreme temperatures in stride, but they must have shelter from rain, snow, and wind.

This simple shed meets that basic need, with the addition of a few welcome amenities. Goats like a shelf for sleeping and lounging. They prefer their food dry and off the ground, so a manger is handy—and reduces wasted hay. They like a large entrance (door not necessary) and enjoy a window to keep an eye on things.

The shed shown here is about as basic as you can get. Its 6 × 8-ft. (1.8 x 2.4m) footprint and modest height (6 ft. at the front, 4 ft. [1.2m] at the back) make it backyard friendly. The ½-in. (12mm) exterior plywood sheathing works fine as the finished siding or can be the basis for lap or board-and-batten siding if you want something more stylish.

Layout is the only challenging aspect of this project. For strength, the plywood overlaps the floor framing and helps solidly join the corners. That means that the sheathing extends 4 in. (10.2cm) beyond the bottom of each wall. In addition, the front and back wall sheathing overlaps the sidewalls to solidly join the corners. You'll also need to plan your framing carefully so that plywood joints are fully supported.

Protection from rain, snow, and the prevailing wind are all that goats ask for. This simple shed, above, gives them a dry shelter open enough so that they can keep an eye on things. The extended roof protects the manger, which is set at browsing height.

Goats like to keep their feet dry and feel most secure perched somewhere. A simple sleeping shelf, left, meets both needs. A window is an optional addition, though goats seem to appreciate it.

Tools		Materials		
Stakes and mason's line	Pry bar	Standard concrete blocks (for	2 10-ft. (3m) 2x6s	exterior screws
Circular saw with a	Hammer	foundation)	26 8-ft. (2.4m) 2x4s	Roll roofing
12-24-tooth	Handsaw	8 4 x 8-ft. (1.2 x 2.4m)	8 10-ft. (3m) 2x4s	Drip cap
carbide tip blade	Drywall T-square	sheets of exterior-	4 12-ft. (3.7m) 2x4s	Roofing nails
Saber saw	Speed square	grade ½-in. plywood	16d galvanized	Roofing sealant
Sawhorses	Framing square	2 4 x 8-ft. (1.2 x 2.4m)	common nails	Acrylic sheet for
Measuring tape	Squeeze clamps	sheets of pressure-	or 3-in. (76mm)	window
Cordless drill-driver	Utility knife	treated ¾-in. (1.9cm)	exterior screws	4 ½-in. (13mm)
Drill and driver bits		plywood (for the	8d galvanized	pan-head
		floor)	common nails	screws
			or 1½-in. (38mm)	

Goat-Shed Plan

This diagram provides the essential dimensions you'll need to frame your shed. It also guides you in cutting and positioning the plywood to ensure sound joints and efficient use of material. Note that the floor is framed with joists 12 in. (30.5cm) on center (O.C.) for strength. Walls are 24 in. (61cm) O.C., and the roof rafters are 16 in. (40.6cm) O.C.

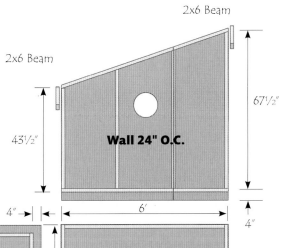

More than decoration, the beams simplify roof construction, allowing you to frame the roof on the ground, then slide it onto the shed without worrying about lining up rafters with wall studs.

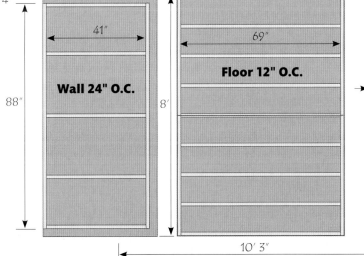

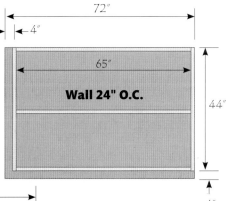

If you are new to carpentry, use exterior screws as fasteners. They cost a bit more and are slower to install, but you can back them out if you make a mistake. They also hold extremely well.

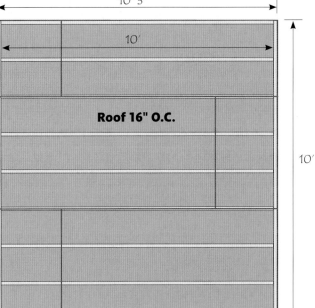

Goat Shed

1

Frame the floor so that the outside perimeter is 72 x 96 in. (182.9 x 243.8cm). Position foundation blocks using stakes and mason's line. Place the 2x4 joists 12 in. (30.5cm) O.C., fastening with 16d galvanized common nails (or 3-in. [76mm] exterior screws).

2

Check the frame for level, adjusting the blocks as needed. Check all four sides. Be as precise as you can, but be aware that some settling will be inevitable.

3

Square the floor using a full piece of plywood—the factory edges are as good a guide as you can find. Mark the plywood for fastening to the joists (inset).

4

Fasten the plywood to the floor using 8d galvanized common nails (or 1½-in. [38mm] exterior screws) every 10 to 12 in. (25.4 to 30.5cm). Cut the plywood to size, either on sawhorses (supported by 2×4s to avoid binding) or after you have attached the plywood. (See Step 12.)

5 Lay out 24-in. (61cm)-O.C. studs for the angled wall frame on the 2x4 soleplate. Next, transfer the stud locations to a 1×4 of equal length. Important: make sure one stud is centered 48 in. (121.9cm) from one side so that you can apply a full sheet of plywood. Cut two wall studs, the longer at 72 in. (182.9cm) and the shorter at 48 in. (121.9cm).

Getting Kids Involved

Backyard-homestead construction is a great way to get children involved and introduce them to carpentry basics. Hammering nails is a good basic. (Hold the hammer low on the handle, and swing from the shoulder.) Using a handsaw is an empowering skill. (Let the saw do the work, and use the full length of the blade.) Measuring skills are less exciting but introduce inches and feet. (Make dimension marks using a "V" for precision.) Most of all, just being part of the project helps kids gain intuitive knowledge about how things to go together.

6 Fasten the 72- and 48-in. (182.9 and 121.9cm) studs on each end of the soleplate using 3-in. (76mm) exterior screws. Fasten the marked 1x4 36 in. (91.4cm) up the wall using 1½-in. (38mm) screws. Attach the middle studs as shown, making sure that each runs a bit long. Align them with the marks on the 1x4, and fasten them to it. Mark the longest stud at 67½ in. (171.5cm) and the shortest at 43½ in. (110.5cm) Mark along a 2x4 to capture the angle.

7 Set the circular saw to match the angle. Make a test cut on a scrap piece of 2x4; check its accuracy; and adjust as needed. Number and unfasten each stud.

Continued

Goat Shed (cont'd)

8

Make the angled cuts using a speed square as a cutting guide. Secure the stud to sawhorses with clamps before cutting.

9

Assemble the wall frame, reattaching the 1x4 as a guide to lining up the studs accurately. Square up the frame using a framing square. Without trimming the top plate to size, fasten it in place using 16d galvanized common nails or 3-in. (76mm) screws.

10

Use a handsaw to trim both ends of the top plate. Build a mirror-image frame for the opposite wall, repeating the same process.

11

Measure 4 in. up from one end of a sheet of plywood. Tack two nails on the mark to act as stops. Place the plywood along the edge of the frame so that the tacked nails hit against the soleplate. Re-measure the overhang.

12 Strike a cut line using a chalk line or straightedge, and trim the plywood. Have a helper support the scrap as you near the end of the cut and set it aside for sheathing the roof.

13 Rip a sheet of plywood lengthwise using a straightedge as a guide. Always be sure to support plywood with four 2x4s to avoid binding when you cut.

14 Finish sheathing the sidewall, applying a fastener every 8 to 10 in. (20.3 to 25.4cm). Sheath the opposite wall. Frame and sheath the rear wall, following the diagram on page 195.

15 Raise one wall, carefully lining up it up with the front edge of the floor. Screw or tack the soleplate to the flooring using 16d galvanized common nails or 3-in. (76mm) screws.

Continued

Goat Shed (cont'd)

16

Raise the back wall, carefully sliding it firmly up against the sidewall. Align the corners , and fasten from the outside using 8d galvanized nails or 1½-in. (38mm) exterior screws. Push it firmly in place against the floor, and fasten the bottom.

17

Check for plumb, and finish fastening the two walls. Use two fasteners every couple of feet along the soleplate and one every 12 in. (30.5cm) where the plywood overlaps the corner. The way the corners come together squares up and strengthens the shed.

18

Erect the remaining sidewall and the front wall. If you want a window in the remaining side, cut it before raising the wall using the method shown on page 182.

19

Cut two 10-ft. (3m) 2x6 beams to support the roof at the front and back of the shed. For a decorative look, make a mark 3 in. (7.6cm) up from the bottom on each end and another mark 5 in. (12.7cm) in from each end on the bottom. Strike a line between the marks, and cut the ends as shown.

20

Attach one of the beams to the front of the shed by making a mark on the beam 12 in. (30.5cm) in from each end. Attach the beam using a single 3-in. (76mm) screw on one corner so that it is 1¼ in. (32mm) above the top plate. Level the beam while a helper checks that the sidewall is plumb. When the beam and wall look right, fasten the board using two fasteners in every stud. Mount the second beam on the back of the shed, about ¼ in. (6.4mm) below the top plate.

21

Frame the roof using 10-ft. (3m) 2x4s as the front and back pieces and 8-ft. (2.4m) 2x4s as the rafters, fastening with 16d nails or 3-in. (76mm) screws. Attach the rafters 16 in. (40.6cm) O.C.

22

Square up the roof as you apply one 4 x 8-ft. (1.2 x 2.4m) piece of plywood. Nail the plywood using 8d nails or 1½-in. (38mm) screws every 8 to 10 in. (20.3 to 25.4cm). Do not apply any additional sheathing at this point.

Continued

Goat Shed (cont'd)

23

Tip the roof onto the shed, orienting the rafters as shown. Slide it in place, lining up its edges with the ends of the beams and with at least 2 ft. (61cm) of overhang in the front (inset), enough to keeping precipitation from blowing inside the shed. That will leave about 1 ft. (30.5cm) in the back.

24

Fasten every other rafter using brackets as shown or by toenailing with 16d nails or 3-in. (76mm) screws. Apply sheathing, being careful to stagger the seams and using 8d cement-coated nails every 8 to 10 in. (20.3 to 25.4cm).

25

Add drip-edge flashing over roofing felt, or tar paper, if you choose to use it, at the top eave and sides and under the felt at the bottom eave. Staple the roofing felt if you plan to roof immediately. Otherwise, tack down battens to keep the paper from blowing off.

26

Add roll roofing following the steps shown on page 206. Cutting the roofing on the ground makes the job easier. Alternatively, put down a layer of roofing felt, and add corrugated aluminum roofing.

Not Up to Stick-Building?

If your schedule or skill level doesn't permit you to build the goat shed from scratch, consider purchasing ready-made or kit sheds. The 6 x 8-ft. (1.8 x 2.4m) shed shown below costs about $1,200 and can be shipped anywhere in the continental U.S. as a kit. Once it arrives in knocked-down form, assembly requires only modest carpentry skills. Perimeter 6×6s bolt together with large angle irons. All four walls are complete, needing only to be tipped up on the 6x6 frame and fastened together. Rafters arrive in truss form. Roof sheathing is precut and ready to nail in place. Even shingles are included. The shed is available from Eagle Sheds (www.eaglesheds-andgazebos.com).

Ready-made sheds are fabricated in an assembly-line manner that cuts costs and may offer better quality than on-site, stick-built sheds. The completed shed is trucked to your site and set in place. Most are regional due to the prohibitive cost of shipping long distances.

27

Add the goat shelf starting with a simple 2x4 frame 2 ft. (61cm) wide and long enough to attach to the wall studs. Position it 16 in. (40.6cm) off the floor. Add crosspieces every 24 in. (61cm) and a central support leg (inset).

Continued

Goat Shed (cont'd)

28

Top off the shelf with ½-in. (12mm) plywood notched to set into the sidewalls so that there is no gap. This tight fit is important because serious injury could result if a goat caught its foot in a crevice.

Fence Feeding

In the spirit of keeping things off the ground, a fence can come in handy too. For all-stock sweet feed (a cup or so a day), and salt (a constant requirement, inset) attach these handy holders to a fence—or to the inside of the shed.

29

Add a backboard roughly 2 ft. (61cm) high as an additional safeguard against getting a foot caught. Both the backboard and shelf are good places to use up leftover plywood scraps.

30

Attach an acrylic sheet to windows. Cut it to size using a straightedge and utility knife. (See page 140.) Predrill holes, and fasten with four ½-in. (13mm) pan-head screws.

Make a Manger

Goats are browsers and prefer to reach for their food. They are also fastidious and won't eat hay from the ground. That's why a manger is a good addition to your shed. This manger is 41 in. (104.1cm) wide and 27½ in. (69.9cm) high, but you can flex the basic design to suit your herd. The plywood backboard simplifies mounting— apply fasteners through the plywood anywhere you can locate framing members in the wall. In addition to scraps of ½-in. (13mm) plywood for the back and sides, you'll need a 12-ft. (3.7m) 2x2 and two 8-ft. (2.4m) 1x3s.

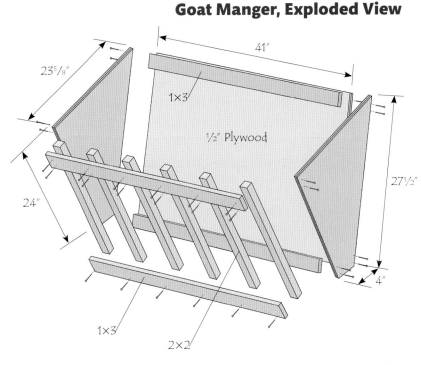

Goat Manger, Exploded View

41"

23⁵⁄₈"

1×3

½" Plywood

24"

27½"

4"

1×3

2×2

1. Cut two 1x3s 40 in. long, and mark both pieces for attaching them to six 24-in. (61cm) 2x2s about every 6⅜ in. (16.2cm). Drill pilot holes, and fasten using 2½-in. (64mm) exterior screws.

2. Cut two sidepieces, following the diagram at the top of the page. Cut a backboard 41 x 27½ in. (104.1 x 69.9cm). Add 40-in. (101.6cm) 1x3 nailers to the backboard, and then attach the sides (inset).

3. Complete the manger by attaching the front grill from Step 1 using 2½-in. (64mm) exterior screws.

Roofing Alternatives

Asphalt shingles (pages 168–169) are the most common roofing option, but roll roofing, wood shingles, and metal roofing are also great materials for backyard sheds. Roll roofing is the least expensive and simplest to install. Unlike shingles, which need a 4-in-12 slope, roll roofing suits the slightly inclined roofs common on sheds. And though it won't win any awards for looks, it does fine on more steeply pitched roofs. On the downside, it won't last as long as the other kinds of roofing shown in this book.

Tools	Materials
Chalk-line box	Roofing felt
Utility knife	Roll roofing
Flat pry bar	Hot-dipped galva-
Broom	nized roofing nails
Hammer	Roofing cement
Trowel	
Applicator brush	

Roll Roofing

1

Roll out the first layer of roll roofing (inset), following a chalk line snapped 35½ in. (90.2cm) from the eaves, and nail at 12-in. (30.5cm) intervals. Spread roofing cement on the first course.

2

Roll out the second layer, nailing every 12 in. (30.5cm) with roofing nails. The doubled first course guards against moisture being driven under the roofing.

3

Strike a chalk line as a guide for placing subsequent courses of roofing. Overlap a minimum of 4 in. (10.2cm). Trowel on roof cement where the course overlaps.

4

Successive courses cover the roof cement. Cement and nail vertical seams (inset) as you did the horizontal laps. Apply roof cement, and nail along the rake.

Wood Shingles

Wood shingles and shakes are usually made from western red cedar. Shakes are thicker and more durable than shingles, often lasting 40 years—twice as long as shingles. Both shakes and shingles are graded #1–#3, clear to knotty. Wood shingles are not recommended for roofs with less than a 3-in-12 slope; shakes for a slope of less than 4-in-12. With a 3-in-12 slope, 16-in.-long (40.6cm) shingles may have a maximum 3¾-in. (9.5cm) exposure (5 in. [12.7cm] for 4-in-12); 18-in. shingles (45.7cm), a maximum of 4¼ in. (10.8cm) (5½ in. [14cm] for 4-in-12); and 24-inches, 5¾ in. (14.6cm) (7½ in. [19.1cm] for 4-in-12).

Tools	Materials
Hammer	4d or 6d galvanized
Carpenter's pencil	box nails
Utility knife	Heavy-duty staples
Block plane	Plastic mesh
Saber saw	Drip edge
Spacing jig	Shingles or shakes
Staple gun	

Wood Shingles

Prep the roof using felt paper, drip cap, and plastic mesh for air circulation under the shingles. Alternatively, apply horizontal skip sheathing to the shed rafters (inset).

Install a starter course two shingles thick. Extend beyond the drip cap 1 in. (2.5cm). Overlap shingle gaps by 1 in. (2.5cm) TIP: To avoid splits, tap the point of the nail with a hammer.

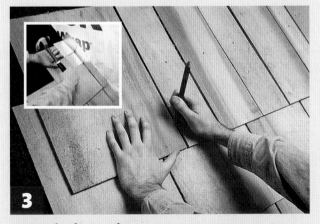

Keep ¼ in. (6.4mm) between singles, using a pencil as a guide. Cover every nailhead by at least 1 in. (2.5cm) of the next shingle course. Use step flashing (inset) where the roof

Use a spacing jig to keep the exposure consistent. To trim a shingle lengthwise, score it several times with a utility knife, and split it. Use a block plane to smooth it.

Installing Metal Roofing

Standing-seam metal roofing can cost three times more than composite shingles but lasts up to 50 years almost maintenance-free. Metal roofing can cover roofs that have a slope of at least 3-in-12; some metal roof systems can handle slopes as slight as ¼-in-12. Standing-seam panels run vertically and interlock at the seams. They are made from aluminum or galvanized steel in a variety of finishes. Panels can be flat or ribbed between seams and ordered in lengths up to 40 ft. (12.2m).

Installation involves laying 12- to 16½-in. (30.5 to 41.9cm)-wide panels and joining them at the seams, wall flashing, and ridges. The manufacturer cuts the panels to the length you order; consult its installation guide for how to take measurements. You can apply metal roofing over plywood covered by 30-pound felt. Never step on metal roofing—it dents, scratches, and crimps easily—and wear work gloves when handling it.

Tools	Materials
Measuring tape	Metal roofing panels
Cordless drill-driver	Flashing
Bits, including hex-nut driver	Sealant or caulk
Hammer	30-lb. roofing felt
Framing square	Fasteners
Aviation shears (two pairs)	
Pry bar	
Caulking gun	

Standing Seam

Some panel seams are not preformed but joined on site using a machine that travels on wheels along one seam at a time, folding the sections together. The simplest DIY systems have panels without mounting clips. Instead, you fasten each panel directly to the roof through a nailing flange, snapping down the successive panel onto this seam on one side and fastening with nails on the other as you work across the roof.

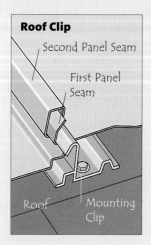

Roof Clip
Second Panel Seam
First Panel Seam
Roof
Mounting Clip

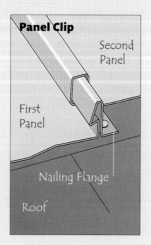

Panel Clip
Second Panel
First Panel
Nailing Flange
Roof

Corrugated Panels

Corrugated aluminum and galvanized-steel panels are long-lasting solutions for utility buildings; plastic and fiberglass corrugated panels can provide a watertight yet translucent covering for chicken runs, sheds, and greenhouses. Panels of both types are typically sold along with manufacturer-specific nails, filler strips, and caulk. Because these panels do not interlock like standing-seam panels, you must face-nail them. To prevent leaks, every fastener has a rubberized washer. The trick is to set the fasteners just firmly enough to seat the washer without deforming it and causing a leak.

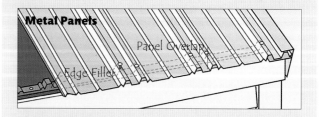

Metal Panels
Panel Overlap
Edge Filler

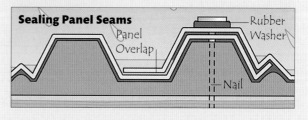

Sealing Panel Seams
Panel Overlap
Rubber Washer
Nail

Metal Roofing

1

Begin by installing the eave flashing over 30-lb. felt. Use wood screws as specified by the manufacturer. Add a bead of caulk or two before adding panels (inset).

2

Install the first panel so that it overhangs the eave flashing according to specs. Check that you have oriented the panel correctly so you can clip on subsequent panels.

3

Clip the next panel in place by pressing down evenly with your palm. Panels can crimp, so enlist a helper. Add fasteners to secure the panel. Work your way across the roof.

4

To cut a panel without damaging it, use two shears simultaneously. You'll likely have to rip the final panel on your shed and around walls and obstructions such as cupolas.

5

Install rake flashing, applying sealant and approved fasteners, typically hex-head screws with rubber washers.

6

Seal the top edge of the roof with ridge cap for a gabled roof or suitable flashing for a shed-type roof.

Setting Up a Backyard-Homestead Shop

Building the projects in this book is much easier if you have a sheltered, all-purpose shop. Whether it is a corner of the garage or something grander, you won't regret having an organized workplace to get projects done well and quickly. A basement shop is the least convenient because it is hard to trundle large projects up and down stairs.

Access is important, especially if you build coops, gates, or beehives in the winter. Installing double doors on sheds makes sense. Strategically-placed windows help, too: a window located at the right height near a power miter saw lets you run long stock out the window while cutting. (See the illustration opposite.)

Power and Light. A dedicated circuit is ideal for a shop, but because you are unlikely to be running more than one power tool at once, you can get by with a shared 120-volt circuit. The circuit should be equipped with a ground-fault-circuit-interrupter (GFCI) breaker or receptacle. Plan for a receptacle every couple of feet along your bench and every 6 feet along the wall. Utility fluorescent fixtures are the best bet for illumination. When mounting them over a table saw or bench, position them so that you won't be working in your own shadow.

Storage. Make the best of overhead space using racks or hangers. Scraps—essential to any shop—need organizing, too. Bins, buckets, or boxes make it easy to see what you have got.

Tools. Typically, the more you pay for a tool, the longer it will last and the more you will enjoy using it. It makes sense, however, to economize on tools that you you will seldom use. A set of chisels can be expensive, for example, but you will rarely need them for backyard-homesteading jobs; on the other hand, you will use a cordless drill-driver almost constantly. Buy one rated at 18 volts with a ⅜- or ½-in. (10 or 12mm) chuck. You can cut almost anything with a circular saw and may want to hold off on the convenience of a table saw and a power miter saw. Or if you are wary of power saws, start with a saber saw. It is slower and requires a guide for straight cuts, but it is less daunting than a circular saw.

Tools for the Shop

Table saw	1-, ½-, ¼-in. (25, 13, 6mm)
Power miter saw	Butt chisels
Two-wheel grinder	Cordless drill-driver
Drill press	(plus twist and driver bits)
Extension cords	Carpenter's level
Eye and ear protection	Claw hammer
Respirator	Squeeze-type stapler
Gloves	Adjustable wrench
Measuring tape (least	Philips screwdriver
two in case you	Standard screwdriver
misplace one)	Spring clamps
Speed square or	C-clamps
combination square	Squeeze clamps
Framing square	Caulk gun
Chalk-line box	Cutting pliers
Utility knife (several)	Wire-stripper pliers
Cross-cut and ripping	Locking pliers
saws	Lineman's pliers
Sawhorses	Pry bar
Circular saw	Wrecking bar
Saber saw	Hacksaw
Bow saw	Sledge hammer
Block plane	Rubber mallet
Perforated rasp	Wire brush
Sanding block	Come-along ratchet
Flat double-cut file	and cables

Supplies

3-in. (76mm) deck screws	Waterproof adhesive
2½-in. (64mm) deck screws	Sealant
1¼-in. (32mm) deck screws	Paint
16d common nails	Wood scraps
10d common nails	Bits of tubing, hose,
8d common nails	wire, screen
6d common nails	Silicone spray
4d common nails	Oil

Backyard-Homestead Workshop

A strategically placed window lets you cut long stock.

A woodworking vise grips things that otherwise might be marred by a steel bench vise.

Old kitchen cabinets with the doors removed make great shelves.

Choose simple multipurpose hangers for perforated hardboard.

A two-wheel grinder makes it easy to keep gardening tools sharp.

2x4s supported by ropes are great at getting lumber up and away.

Simple ½-in. plywood supports fasten easily to wall studs and support a surprising amount of lumber.

A simple table for your power miter saw makes for speedy, accurate cuts.

Three buckets bolted together organize wood scraps.

Build your bench strong and heavy. Use 4x4s or doubled 2x4s for legs and a couple of layers of particleboard for the top. A second bench in the middle of the shop is handy.

Surprising Tools

A few tools would not be part of a home workshop but are useful to the the backyard homesteader.

- **Come-Along.** For moving a chicken coop, removing boulders, stretching fence—any number of chores where pulling strength is needed—a come-along is indispensible. A lever and ratchet produces more power than an entire tug-of-war team.

- **Drywall T-Square.** For laying out projects on sheet goods and as a saw guide, this tool is well worth the investment.

- **Locking Pliers.** While this tool is not known for its finesse, it pulls nails, grabs bolts, and levers pipe with the best of them.

- **Hose Shears.** A utility knife cuts tubing and hose but not half as neatly and quickly as this tool.

- **Pocket Knife.** You should always have sharp knife with a blade of at least 4 in. (10.2cm) in your pocket when doing chores. Just don't take it to the airport.

- **Baby Sledgehammer.** A standard 16-ounce hammer will make a mess of stakes or poles when you drive them. A baby sledge whacks them just right.

CHAPTER 5

Wind and Solar Power

Installing a Pump or Aeration Windmill

Using a windmill to pump water dates back at least 2,000 years, and famously has kept the sea out of the Netherlands for centuries. In the United States, ranch and farm use predates electricity and for many remote areas remains the best way to pump water for livestock. The classic rural windmill has been pumping for 120 years. It uses a 6-ft. (1.8m) or larger fan to power a gearbox that creates the pumping suction to pull water from a well. (See page 214.)

That kind of windmill was, and still is, the best way to pump a lot of water from a well. A new variation on the windmill theme, however, uses a compressor in place of the gearbox. By means of a tube running down the tower, it powers a submersible air-drive water pump capable of lifting water from wells up to 40 ft. (12.2m) deep. It can also lift water from a pond or brook to fill a storage tank for drip irrigation or to aerate the pond. The advantage of the compressor-type windmill is cost: a small compressor windmill costs about $1,500, while a new gearbox-type is more than $4,000.

They do require wind, however, a factor well worth checking. Your local airport can give you typical wind direction and speed for your area. If you are near a large city, the data on the National Oceanic and Atmospheric Administration (NOAA) site may be helpful. You should locate your windmill 150 feet away from any obstructions in the path of the prevailing wind. Have the average local wind speed and the amount of water you hope to pump or aerate available when contacting manufacturers. If you occasionally have winds of more than 25 mph, you'll be reassured to know that standard equipment is an over-speed mechanism that rotates the fan out of the wind direction.

The project that follows shows the installation of a tower and fan powering a compressor linked to an aerator. If you install one yourself, plan on pulling together a crew to help erect the tower, barn-raising style.

An icon of rural life, a windmill can be a handsome addition to a suburban or rural backyard homestead. It can be equipped with a well pump or a pond aerator.

Tools	Materials
Posthole digger	Windmill kit
Shovel	Pier forms
Hammer	Concrete mix
Cordless drill-driver	Post anchors and
and bits	fasteners
Socket wrench set	
Adjustable wrench	
Measuring tape	
Sawhorses	

214 Installng a Pump or Aeration Windmill

218 Installing Solar Power

222 Wind Turbines for Electricity

224 How Wind Systems Work

225 Hybrid Systems

213

5 Wind and Solar Power

Windmill Perspective View

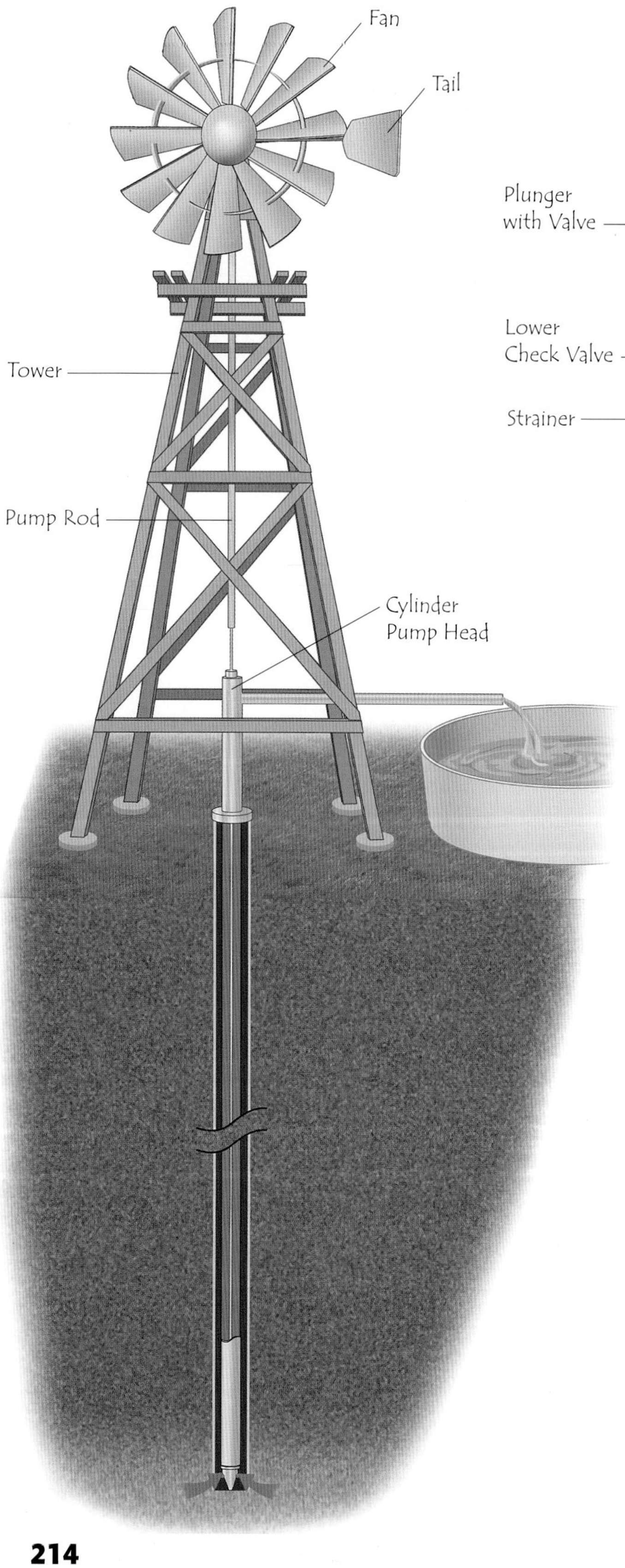

Fan

Tail

Tower

Pump Rod

Cylinder
Pump Head

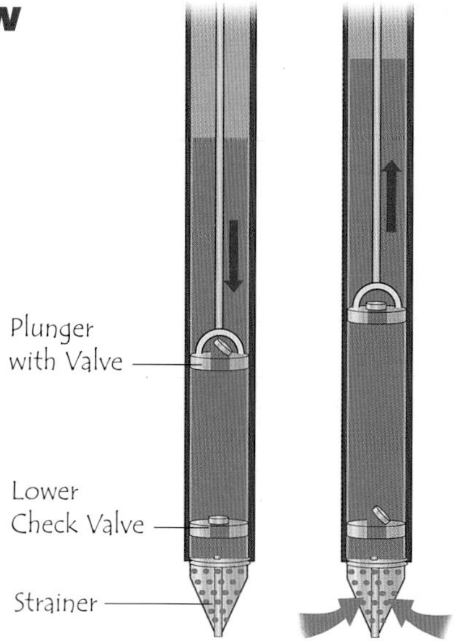

Cylinder Pump

A cylinder pump raises a column of water in installments with each upward pull of the pump rod. As the plunger descends, a valve opens to let water flow in above it. On the upstroke, that valve closes and water is lifted up the well pipe. At the same time, more water is pulled in through the strainer. The result: a column of water is raised out of the ground.

Plunger
with Valve

Lower
Check Valve

Strainer

Gracing the rural horizon for more than 100 years, the windmill still is hard at work pumping water and aerating ponds. This diagram shows the traditional gearbox version that converts rotary power to reciprocal motion, which raises and lowers a cylinder pump deep in the well to effectively pull up a column of water.

The rod that runs inside the tower has been replaced by a hose in newer versions such as the one shown in the steps that follow. It uses a compressor connected to the fan to power a pump or provide pond aeration.

Which Windmill Is Right for You?

For pumping water from a well as deep as 120 ft. (36.6m), choose the traditional gearbox-type windmill. A 6-ft. (1.8m) fan on a 21-ft. (6.4m) tower will pump 180 gallons of water per hour in a 15–20 mph wind. Cost: about $4,000 and up.

For pumping water from a well 40 ft. (12.2m) deep or less, choose a compressor-type windmill. A 6-ft. (1.8m) fan on a 16-ft. (4.9m) tower will keep a 500–1,000-gallon stock tank full or aerate a small pond. Cost: About $1,500. Compressors can be equipped with valves so that you can switch between aeration and pumping.

Installing a Pump or Aeration Windmill

Pour piers, and set post anchors, carefully following the measurements provided by the manufacturer.

Lay out the components of your windmill kit. With most kits, the components are marked. Position the marks for easy reference as you grab them.

Splice the 4x4 legs together using angle iron provided with the kit. Use a straight 2x4 as a guide as you insert and fasten the lag screws.

Assemble two sides of the tower, positioning the legs so that the angle-iron bracing is oriented to the inside of the tower.

Continued

Installing a Pump or Aeration Windmill (cont'd)

5

Set the completed sides next to each other, and complete the tower. Join the two sides at the top; then work your way down, attaching cross members as you go. Add the upper platform, and install the metal plate that holds the fan assembly.

6

Carry the tower to its final location before adding the fan. Orient it so that your team can readily lift it onto its anchors.

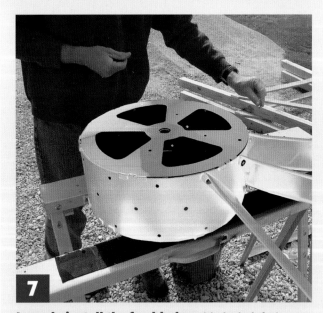

7

Loosely install the fan blades with the hub facing down. Insert the bolts and washers provided, fastening the nuts from the inside of the hub. Follow the manufacturer's instructions closely to avoid damaging the blades.

8

Add the fan-blade bracing between the blades, again keeping the fasteners loose. Once you have installed all of the braces, flip the hub over and tighten all of the nuts on the blades and blade braces. Assemble the tail.

9

Insert the pivot tube, and attach the compressor. Attach the blades and tail. Complete all connections specified by the manufacturer. Raise the tower onto the piers (inset). Have plenty of people on hand, some lifting, and some pulling down on the base of the tower, and some providing steady pressure with a rope or two.

10

Immediately fasten the tower to the anchors using exterior screws. Complete the installation of aerator or pump linkage.

Drill Your Own Well?

Hiring a well driller will cost you $12–$15 a foot and is often your only option if you need a deep well or have rocky soil. However, if you need an irrigation well and can reach water about 25 ft. (7.6m) down, here's a do-it-yourself option worth considering. Much like the method shown on page 256 for boring under a sidewalk, it uses water to sink PVC pipe into the ground. Key to the process is a homemade PVC drill head fed by two garden hoses. The business end of the PVC pipe is toothed to help chew into the soil. Drilling debris is pushed to the surface by water pressure from the hoses. As you work the pipe into the ground, you can attach additional sections to extend the depth of the well. Visit **http://drillyourownwell.com/links.htm** for more details.

Water from two hoses provides enough force to raise the cuttings as the PVC pipe slowly works its way down.

Fabricated from PVC components available at any home center, the drill head is the key component of the rig.

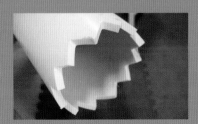

Use a handsaw or bench grinder to serrate the business end of the well drill.

An oak plank, U-bolts, and hose clamps make the yoke for swiveling the rig back and forth as you drill.

Installing Solar Power

On freezing nights, this system powers a 100-watt heater for a chicken water dispenser and supplemental lighting to keep up egg production. During the day, it powers the refrigerator and freezer—as well as a few other household appliances.

The system, which is installed in rural Illinois, uses five 200-watt panels attached to the garage, where the charge control, batteries, and inverter are also installed.

Here's how the system works: On a sunny day, each solar panel produces 200 watts of 12-volt electricity. The power runs through a charge control, which keeps the system from overloading the storage batteries—12-volt deep-cycle batteries that store the power until needed. When a device called a load control draws electricity, it pulls the power through an inverter, transforming the power from 12-volt DC to 120-volt AC—residential voltage.

Although a photovoltaic system like this saves you the trouble of running underground lines and offers the prospect of free electricity, it is not cheap. The system shown cost $2,300—an exceptionally low price because the solar panels were bought secondhand.

Note: These steps and components represent a typical solar installation. Your choice of solar panels, the number of panels, and the capacity of your system may, however, require components different from the ones shown. Follow the manufacturer's instructions.

Tools	Materials
Cordless drill-driver and bits	Solar panels
Hole saw	Panel leads
Adjustable wrench	Splicing hardware
Socket wrench set	Brackets
Cutting pliers	Bracket clips
Needle-nose pliers	Outdoor conduit and fittings
Wire stripper	Electrical cable suited to the voltage and amperage of your panel array
Groove-joint pliers	
Philips screwdriver	
Standard screwdriver	
	Wire connectors
	Electrical tape
	Cable-pulling lubricant
	Anti-corrosive lubricant
	Charge controller
	12-volt batteries
	Nonmetal or coated rack for batteries
	Battery cables
	Construction paper for covering panels
	Automatic transfer switch

How Solar Works

1. When the sun is out, the solar panel converts photons into electrons of direct current (DC). **2.** The charge control cuts the flow of direct current to the batteries when they are charged, avoiding an overload. **3.** 12-volt batteries store the electricity. **4.** The inverter transforms 12-volt direct current (DC) to 120-volt alternating current (AC). **5.** The automatic transfer switch sends excess power back into the power grid—and a credit to your electric bill—when power is not needed for the poultry house and refrigerator/freezer.

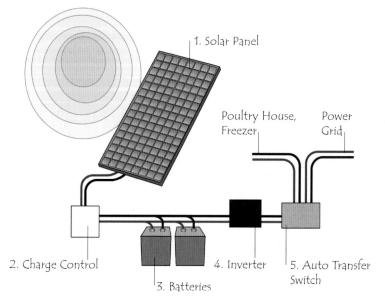

1. Solar Panel

Poultry House, Freezer

Power Grid

2. Charge Control

3. Batteries

4. Inverter

5. Auto Transfer Switch

Installing Solar Power

1

Attach the panel-support brackets to framing—in this case, under the eaves to hold panels without compromising the roofing. Bore holes (inset) for attaching clips.

2

Add the bottom support bars to the brackets. Tilt each attachment clip so that it aligns with the plane of the roof (inset).

3

Add retaining brackets to the panels for attaching to the support bars at the eave and the bottom of the panel. Set the panels in place, and tighten the brackets (inset).

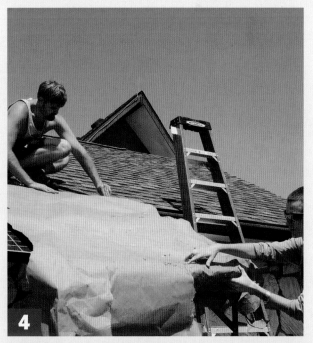

4

IMPORTANT: cover panels with paper or other sunlight-blocking material to avoid the hazard of a shock while attaching positive leads.

Continued

Installing Solar Power (cont'd)

5

Run cable from the panels to the area where you'll install the inverter and batteries. Each panel has positive and negative leads (inset) on its underside.

6

Install LB fittings on the soffit. These fittings allow access for pulling cable through the conduit. They are rated for outdoor use.

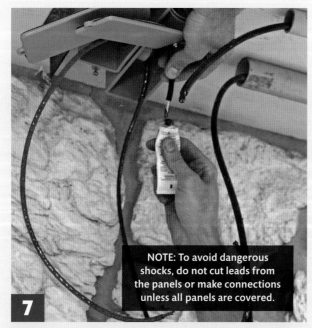

NOTE: To avoid dangerous shocks, do not cut leads from the panels or make connections unless all panels are covered.

7

Pull cable to the LB fittings and through the conduit. Strip cable for connections, and apply anti-corrosive lubricant.

8

Connect the leads using the connectors attached to them. Where there are no connectors, use a splice fitting (inset) and pull the protective insulation over it.

9

Install the charge control box to a stable backboard, and pull cable to it. For long runs you may have to use cable lubricant. Do not use liquid soap, oil, or grease—they can damage the cable insulation.

10

Make connections at the charge control box following the manufacturer's instructions. This important component senses when the batteries are fully charged and cuts power from the panels to avoid overloading.

11

Install the batteries and inverter in a stable nonmetal or plastic-coated rack. Connect the batteries in a series.

12

Install and wire the automatic transfer switch, which routes excess electricity to the power company. Run power from the inverter (inset) to the switch.

Wind Turbines for Electricity

As long as the wind is blowing, a wind turbine can produce power. That means it can be hard at work through the night and in any weather, come rain or shine. For those who choose a hybrid system that combines solar panels with wind generation, a ready source of power is all but guaranteed.

Small, good-quality wind turbine kits can be purchased for around $1,500. This price tag will get you a system suitable for powering an out building with lights, receptacles, and maybe a pump or a poultry water heater. Powering your entire household is another proposition entirely. To do so, you'll need a turbine equipped with a 2–10 kilowatt generator, a tower 30 to 140 ft. (9.1 to 42.7cm) tall, and an investment of $10K and up.

Blowin' the Wind

How do you know if you have enough wind? You can start with a weather app that includes historical wind patterns for your area. State wind maps also help, as can wind data from nearby airports. However, it is your immediate area that really matters. A series of readings with an anemometer (you can buy one for less than $20) will do the job. Much depends on the level of use you'll make of the batteries charged by wind, but if you can't regularly depend on 7–10mph wind, a wind turbine is probably not for you.

Obviously, you'll want to site your turbine away from wind-blocking trees and buildings and on high ground or atop a structure. Ideally, your wind turbine should

The classic horizontal-axis turbine uses an odd number of blades to maintain balance and limit wear. The more blades, the less wind required to get the turbine generating power. Some have as many as 11 blades.

have a 250-ft. (76.2m) radius of unobstructed space around it—no trees, hills, or buildings. A small turbine needs to be about 30 ft. (9.1m) off the ground for best results. To reassure yourself that you're estimating the best tower height, try this simple test: Release a party balloon on a breezy day, and watch for the point at which it is carried away. If your situation is not ideal, take heart. Your turbine will generate electricity, but with less optimal results.

Anticipating Your Power Needs

Size your system's capacity by adding up the wattage demands you'll place on it. For example, four LED light bulbs will require about 40 watts. Add on a well pump, and

you'll need another 500. Run a hammer drill off the system, and you'll be drawing a whopping 1,000 watts. Allow for the fact that usually everything may not be on at once. Kit manufacturers are good at guiding you through this.

Storing the Power

Lead acid, deep-cycle 12-volt DC batteries, often referred to as "stationary" batteries are the go-to type for turbine setups. They are made for the continuous draw of power that a backyard system requires. Avoid "starter" car batteries. They produce higher amps with the oomph to turn over an engine, but won't hold power over the long term. They also degrade with frequent recharging.

Vertical-axis turbines don't need a tail fin because they naturally accept wind from any direction. However, it produces less electricity from a given amount of wind.

How Wind Systems Work

As the turbine blades turn, three-phase alternating current (AC) "wild" power is produced—which is unusable at this point. For storage in a battery, it is converted to direct current (DC) by means of a rectifier. Combined in one unit with the rectifier is a charge controller that protects the battery from overcharging by sending excess power to a "dump load." Marine deep-cycle batteries are used because they tolerate a continuous draw of electricity. At the back end of the system, an inverter converts power to 120-volt AC power—the kind used by pumps, lights, and power tools.

What to Look for in a Kit

Buying a kit is the simplest and often most cost-effective way to successfully tap into wind-generated electricity. If your needs are modest and you can get by on 12-volt truck and auto-type power instead of household 120-volt current, a kit on a par with those used on campers or RVs could do the job for only a few hundred dollars. A step up is a larger system that generates around 1,600 watts—available for about $1,500. There is a dazzling array of wind turbine options out there, many well worth avoiding. As with most products, those made by producers who have been around for a while are your best bet. They are also most likely to have a responsive

Wind Turbine System Components

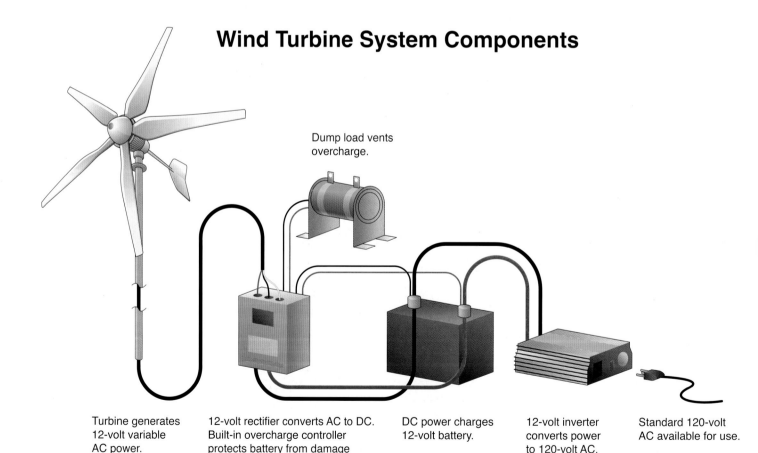

Dump load vents overcharge.

Turbine generates 12-volt variable AC power.

12-volt rectifier converts AC to DC. Built-in overcharge controller protects battery from damage caused by overcharging.

DC power charges 12-volt battery.

12-volt inverter converts power to 120-volt AC.

Standard 120-volt AC available for use.

customer service team to answer your inevitable questions. Be aware that many off-shore manufacturers have jumped on the bandwagon by imitating—with inferior materials—existing systems. Regrettably, many of the systems that pop up on the Internet promise more than they can deliver.

Here are some buying tips:

- Look for blades made with carbon fiber composites or made of high-tensile strength aircraft aluminum. PVC or nylon blades will quickly degrade.
- A horizontal axis turbine (HAT) is more efficient that a vertical axis turbine (VAT).
- A turbine with five or more blades will generate power in lower wind speeds.
- Batteries not included. Ditto the tower.

Hybrid Systems

Variable winds are the norm in most locations, meaning your turbine might not always be turning. And, while solar panels will more than meet your needs in the summer, come the snow and overcast of winter, they will all but shut down. At, night solar panels sleep while a turbine is likely to keep on churning. A system combining a wind turbine with solar panels can help keep your batteries charged. Hybrid kits—including a couple of solar panels, a wind turbine, and a controller that keeps your batteries from overcharging—can help you maintain power even when the weather is not cooperating.

Both systems can be installed by individuals who are handy, willing to follow detailed instructions, and comfortable working at heights. Warning: Do not attempt the final hookup, which is linking the inverter to your electrical panel. Hire a certified electrician.

DIY? Give It a Whirl!

Spinning blades, a generator cranking away, some batteries to store the resulting power: It all sounds simple enough that surely a clever person could cobble together a wind turbine. Perhaps. A web search will yield plenty of DIY schemes, many that are more exercises in wishful thinking than a workable system. Using a salvaged car alternator or an electric motor repurposed to be a generator may seem like a cool, cost-effective approach, but these often fall short. DIY systems require a thorough understanding of the physics of electricity, a delight in tinkering, a well-equipped shop (probably including a welding rig), and a tolerance for less-than-optimal results. Most homeowners will do better with a plug-and-play wind turbine kit. They'll certainly save time. Maybe even money.

Combining wind and solar offers the best of both worlds and all but constant flow of energy.

CHAPTER 6
Aquaponics and Hydroponics

SOIL IS WONDERFUL STUFF, but it also requires a lot of water and amendments, like compost and fertilizer; takes up a lot of space, and can be weed and disease bearing (and well, dirty). Using soil may always be the gardener's first choice, but aquaponics (growing fish and plants together) and hydroponics (growing just plants) offer a viable soil-less alternative for growing fruits and vegetables.

Both systems use of some sort of medium other than soil to hold growing plants. Aquaponic grow beds, for example, commonly use marble-size clay balls or inert rock. Hydroponic systems might substitute plant fiber. Both pump nutrient-laden water to the plants, and both let you harvest produce in a surprisingly short time.

Compared with traditional gardening, where the important inputs come from the soil and the sky, aquaponic and hydroponic systems are highly dependent on human involvement. That means sweating the details and being willing to monitor your system closely.

This chapter is intended as an introduction to both topics, a means of determining whether one might be right for you or not. Both are relatively technical and warrant research: whole books have been written on the subjects and scores of Web sites buzz with forums, tips,

and new ideas. You can also take advantage of DVDs, numerous YouTube videos, and even online courses. To dispel the mystique, you might also want to find someone local who has an aquaponic or hydroponic system so that you can see things firsthand. And as with any new agricultural approach, you'll want to start small before going into serious production.

As a substitute for soil, expanded-clay-based media are used in both aquaponic grow beds (shown) and hydroponic systems. Other grow media include stone made of fired glass, super-heated shale, coco fiber, rock wool, and perlite.

'Ponic Wars

Delve into aquaponics and hydroponics, and you'll soon detect friendly rivalry. Here is how the bragging rights stack up:

Aquaponic pluses
- Feeds plants once an hour, facilitating optimal growth, compared with every 4 to 6 hours for hydroponics
- No nutrients need be added as long as the fish are fed; hydroponics needs nutrient infusions
- No need for an electrical conductivity (EC) meter to see whether nutrient solution must be flushed out occasionally to avoid becoming

toxic—and disposed of properly
- Added dimension of interest provided by fish
- The symbiotic cycle mimics nature

Hydroponic pluses
- Basically plug and propagate, whereas aquaponics requires a 1-month-plus "cycling" startup while nitrifying bacteria get a footing
- Water temperature and pH need not be as closely monitored as with aquaponics
- No worry about fish fighting or dying
- Lower startup costs
- Doesn't require a hefty fish tank

228 Understanding Aquaponics

232 Understanding Hydroponics

233 Assembling a Hydroponic System

Understanding Aquaponics

Avid gardeners turn to compost to fuel their garden. For aquaponic (AP) enthusiasts, it is fish waste that keeps plants flourishing. They may appreciate raising a tilapia or catfish for the table, but it is the stuff that fish leave behind that is the real gold.

And that is what sets AP apart. Aside from power and light, fish food is the only outside ingredient you need to add to the cycle. Here's how it works: Fish waste falls to the bottom of the fish tank and is pumped into a grow bed. The waste-laden water moves through and is filtered by the growing medium—small clay balls or gravel enriched with toxin-scrubbing bacteria. The same bacteria, in concert with casing-producing worms, feed the plants. The filtered water flows from the grow bed back to the fish tank, where the cycle begins again.

What are the advantages?

- You never have to water a plant. In fact, except for the water lost to evaporation and plant expiration, the same water is used repeatedly.
- No weeds invade your garden, which means no cultivation.
- You do not have to compost or fertilize, thanks to the fish waste.
- The table height of the grow beds, necessary to maintain the flow back to the fish tank, makes it comfortable to work your garden.
- It is massively more productive per square foot than a traditional garden.
- Few systems are better if you aspire to four-season gardening, especially if you have a greenhouse.

What can you plant? Almost anything will flourish in an aquaponic setup. The few exceptions are root plants (because the grow medium constrains their growth) and acid-loving plants like blueberries (because AP must maintain a nearly neutral pH).

Building a DIY System

Because of the weight of water (60 gallons weighing more than 500 pounds) and the substantial weight of water-saturated clay or rock grow medium, tanks must be made of stern stuff. Unless reinforced, tanks should be made of ¼-in. (6mm) plastic, minimum. Instead of purpose-built tanks, however, stock tanks, 55-gallon plastic barrels, intermediate bulk containers (IBCs), and reinforced EPDM (ethylene propylene diene monomer) rubber containers fill the bill.

The pump is literally the heart of system and must be rated to move the entire volume of the tank up to the grow bed in 15 minutes. Choose a quality submersible pump—the life of your fish could be at stake if it fails. Look for the following specs: the gallon-per-minute (gpm) pump rate and the head pressure, the gpm when the pump is lifting water up to the height of the grow bed. Buy a reliable timer that can be set to 15-minute increments to keep the flow consistent. A good-quality timer is important because "bargain" timers go only to 30-minute increments.

Plumbing (opposite) is standard PVC and readily available. You can find the bulkhead fittings (to attach the pipes to the tanks) at plumbing-supply or marine stores. The auto-siphon creates the critical pull that draws wastewater through the grow bed. It is carefully engineered and worth buying ready-made online.

AP configurations vary, but the one shown in the illustration at the bottom of this page is one of the better ones. It eliminates the need for a tall fish tank by using a sump tank. In addition, it is easy to add on to. Water is siphoned from the fish tank into the sump, where a pump draws it into the grow bed. On the way, some of the water is directed back to the fish tank for aeration—the all-important process of adding oxygen for the fish. The rest of the wastewater is piped to one end of the grow bed. As the water flows toward the auto-siphon, it provides nutrients for the bacteria and worms in the beds, while the medium filters out solids as the water moves toward the sump. There, the filtered water dilutes the wastewater, and the cycle begins again.

Getting Started

To learn the aquaponic ropes, consider a kit with tank, beds, and pump included. A two-bed system, like the AquaBundance system from Aquaponic Source **(http://theaquaponicsource. com/),** suits a deck or patio and offers 16 sq. ft. (4.9 sq. m) of growing space. It comes with a 60-gallon tank, pump, plumbing, steel stands, and testing gear—almost everything but the grow media, plants, and fish. Cost: About $2,200.

The Aquaponic Cycle

1. Fish are fed up to three times a day, food being the only external input other than light and power. **2.** Wastewater flows into the sump tank. **3.** The pump lifts wastewater mixed with filtered water to the grow bed. **4.** Some water is diverted to aerate the fish tank. **5.** Wastewater flows through the grow bed, providing nutrients for plants. **6.** The auto-siphon returns filtered water to the sump.

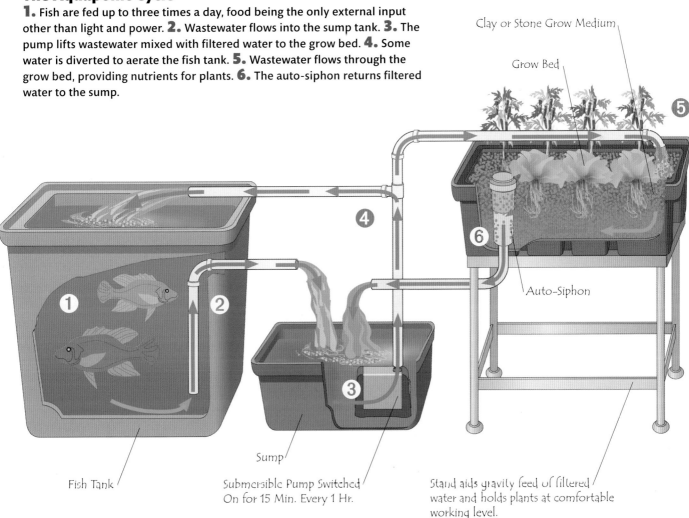

Clay or Stone Grow Medium

Grow Bed

Auto-Siphon

Fish Tank

Sump

Submersible Pump Switched On for 15 Min. Every 1 Hr.

Stand aids gravity feed of filtered water and holds plants at comfortable working level.

Is Aquaponics Right for You?

Space: Any AP system can be run outside during the summer but is practical year-round only in Hawaii and parts of California, Texas, and Florida where warm but not-too-warm conditions prevail. In other climes, the system must be brought indoors during the cold seasons, a sizable undertaking. Some owners run systems in a basement or heated garage. A greenhouse, made to take humidity and spillage in stride, is ideal.

Sound: The splashing of aeration and the pulsing of water as the timer starts the pump takes some getting used to.

Time: Setting up the system takes several weekends, but after that you'll spend about 30 minutes a day on your system. Expect to feed the fish at least once a day. Weekly pH testing is essential; three or four times a week is safest.

Water from the fish tank feeds bacteria and worms living in the grow medium, which in turn produce nutrients for the plants, right. The buildup of nitrifying bacteria is part of the weeks-long "cycling" process that gets an aquaponic system up and running. Bacteria consume ammonia, a waste product from the respiratory process of the fish and toxic to them in high concentration.

Because it is temperature controlled and built for humidity, a greenhouse is an ideal setting for year-round aquaponics, below.

Aptitude: Technophobes may want to stick to traditional gardening. Keeping things in balance takes math skills initially and requires an ability to innovate should something go wrong—like a pump failure. This means monitoring the entire system, something most owners do avidly because they love watching how it works.

Expense: A smallish system with 8 sq. ft. (2.4 sq. m) of growing space costs about $1,400, though if you are handy and willing to scavenge materials, you could cobble a system together for a few hundred dollars. Even once it is up and running, do not expect to beat grocery store prices with your produce or fish. The real benefit is growing pure food with minimal environmental impact.

It's the Water

AP gardeners keep a compulsive eye on their water. You will begin by filling the tank, sump (if you have one), and grow bed with water and running it without fish or plants to let the chlorine dissipate. Add fish to help the system put on filtration muscle by building up nitrifying bacteria—a stage known as cycling. This takes up to six weeks, during which time you'll test pH and make adjustments to bring it to the optimal 6.8-to-7 range that suits bacteria, worms, plants, and fish alike. It takes finesse and experience to know which of a wide range of additives will tweak the pH to where it needs to be. Water temperature is important, too, a good reason why a sheltered setting helps. Fish tolerate a fairly limited temperature range (below): let it get too low, and they won't thrive; too high, and they will die. Monitoring the system, testing the water, and feeding the fish are your regular chores.

Not Just Any Gravel

The grow medium you choose should absorb water while facilitating a flow that spreads nutrients. Not just any gravel will do. Limestone and marble, for example, will dramatically affect the pH. Nor do you want anything that will break down over time. The ideal medium is made of ½- to ¾-in. (1.3 to 1.9cm) chunks, inert, and stable. Expanded clay balls, expanded shale, and glass-based products— among others—fill the bill.

Marble-size fired-clay balls are among of the most popular grow media for aquaponics.

Made of heated and expanded recycled glass, this medium is lightweight and inert.

Fish: To Eat or Not To Eat

If a nice homegrown fish dinner is your first motivation for setting up an aquaponic system, prepare to be patient. It takes even fast-growing tilapia at least 9 months to reach 1½ pounds—a reasonable size for cooking. Trout requires a year of growth—catfish or bass, even longer. The local temperature range is an important guide to which fish to choose. Trout like cooler temperatures, for example, and suffer if the water temperature climbs beyond 70–75 degrees F. Tilapia thrive in water as warm as 90 degrees F. The versatile catfish does well in either extreme.

How many fish can you have? Determine the mature weight of the fish you choose; then allow 5 to 10 gallons of tank capacity per pound of fish. Better yet, determine the size of the grow bed you want; purchase a tank of corresponding volume; and stock the tank with fish accordingly. Where do you get the fish? Buying nonfood fish is simple; just head to your local pet store, and buy freshwater fish of your choice. You can order food fish online from hatcheries and have them shipped to you guaranteed live arrival. Figure about $60 for 25 fingerlings, with an additional $50 shipping.

Tilapia is a warm-water fish.

Trout need cool water and plenty of oxygen.

Catfish tolerate a broad temperature range.

Koi offers good looks but is not a food fish.

6 Aquaponics and Hydroponics

Understanding Hydroponics

Growing produce without dirt is what hydroponic systems are all about. Water conveys all of the needed nutrients, and a medium—pea gravel, floating cups, or coconut mat—supports the plants, dramatically reducing the amount of space needed for a bountiful harvest. The unit shown in the following steps, for example, holds 30 plants but takes up only 6 sq. ft.(1.8 sq. m). The same number of plants in a conventional garden would need a minimum of 120 sq. ft. (36.6 sq. m).—perhaps as much as 300 sq. ft. (91.4 sq. m).. And hydroponics is not only an outdoor option: systems can be set up indoors on tile or stone floors where minor leakage won't be a problem.

Another advantage of hydroponics is growing speed. Because plants are consistently nourished and watered, they come into production faster—as much as 25 percent faster. That can mean up to five growing cycles a year. Hydroponic systems are also virtually weed free and avoid plant damage by many soil-borne pests and diseases.

The challenge of hydroponics is that because no soil is involved, you have to provide everything that the plant needs, which means buying commercially produced nutrients unless you are very good at mixing your own and are willing to gather in bulk all of the needed ingredients. (Organic nutrients are available.) Researching the necessary elements will make you appreciate the complexity of plain old soil.

Hydroponic productivity requires investment of money and time. The kit shown on the following pages costs about $350. Even if you make your own system from scratch, you'll have to purchase items like pans, tubes, wicks, tubing, and pumps. Nor is a hydroponic setup maintenance free. Any system should be checked every couple of days. A configuration with a 5-gallon reservoir will need a new mixture of nutrients every 7–10 days. In hot weather, plants need supplemental water.

Small systems, above, are a great way to determine whether getting into hydroponics in a big way is right for you or not. They are also wonderful for urban backyards and patios (even balconies!) where space is extremely limited.

A hydroponic system is ideal for small lots, doubling if not tripling the productive space at your disposal, below. Because the plants are fed and watered systematically, they grow quickly and produce consistently.

Assembling a Hydroponic System

1

Assemble the frame by pushing the legs into the top portion of the A-frame. As you do so, orient the notches in the legs so that they face outward. Push the legs into the bottom assembly.

2

Slip a tube hanger into each leg until it locks in place. Put the completed frame in a spot that gets a minimum of 6 hrs. of sunlight daily. Check that the frame sits square and flat, with no rocking back and forth or side to side.

3

Position the grow tubes so that the mineral-intake lines—the thin tubing that provides the tube with nutrients—are all on the same side of the frame. Remove the plug, or blind barb, from the end of each intake line. Save the plugs for winter storage.

4

Unpack the pump components. The kit shown includes a pump bucket, submersible pump, timer, manifold (containing feeder lines to which you attach the mineral-intake lines), riser tube, nutrients, and coconut fiber, the medium in which the plants will grow.

Continued

Assembling a Hydroponic System (cont'd)

5

Thread the manifold onto the riser tube. Place the lid on the bucket, and push in the riser tube.

6

Remove the lid, and mount the pump onto the bottom of the riser tube. Replace the lid, and push down on the manifold until you feel the pump hit the bottom of the bucket.

7

Attach the timer to the S-hook provided, and hang it in the hole in the bucket lid. Thread the pump and timer cords through a cable tie, and hang them on the S-hook.

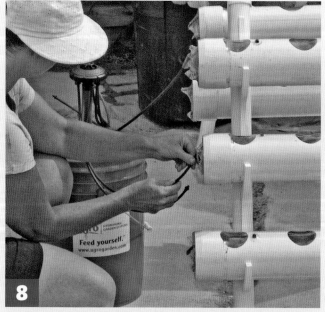

8

Attach the feeder lines to each of the mineral-intake lines. Mix the mineral concentrates following the manufacturer's directions. Fill the pump bucket with water, and add the recommended measures of concentrate, mixing thoroughly.

9

To propagate plant sets: fill the grow tube with coconut fiber, soaking it well; move the feeding drip in the chosen port to one side; use a trowel to form a hole for the plant; then place the plant, completely covering the roots. Put vines in the bottom grow tubes.

Planting Seeds in a Hydroponic System

If you prefer to plant seeds rather than sets, begin by hydrating each grow tube until the coconut fiber is completely soaked. In each port, push the feeding drip to one side, and make a small indentation with a spoon. Add the seeds, and completely cover them. Turn on the system, and confirm that all of the feeder lines are working. Expect quick germination.

10

Plug the timer into an extension cord attached to a GFCI (ground-fault-circuit-interrupter) receptacle. Turn on the system, and run it for 1 min.—one feeding cycle. Check that all feeder lines are working. The small hole beneath each plant should drip slightly. Set the timer for three 1-min. feedings per day.

Building a Langstroth Beehive

Before Lorenzo Langstroth invented his beehive in 1851 (patented in 1852), beekeepers used straw hives—woven "upside-down baskets"—often seen in children's books. Called skeps, they had to be broken open to get at the honey. Langstroth knew there had to be a better way.

His solution was a hive made of stackable boxes that held frames in which honeybees could make their combs. It struck a neat balance between the needs of the honeybee and the convenience of the beekeeper. Now the honey could be harvested without chopping apart the bees' home. In addition, by stacking additional boxes (called supers), the Langstroth hive grew with the colony. His invention soon became universally accepted, used the world over. Now components are easy to find, relatively affordable, and interchangeable.

An assembled beehive like the one shown in the following pages would set you back about $130 if purchased ready-made. You'll need access to a table saw to make it, but otherwise fabrication is simple. Allow one-half day to make the hive. The project begins with the super, the heart of the hive. As you build it, check its dimensions against the size of the frames you intended to use. (See "It Makes Sense to Buy Frames," page 239.)

Langstroth beehives are easy to work with and readily expandable, above. Supers suit three different frame sizes—the smaller the frame, the lighter the super. Sheet aluminum (shown) or a small piece of roll roofing makes a serviceable roof.

The essential components of a Langstroth hive include the outer cover (roof), inner cover, bottom board (the bee entrance), and box-like super, left. The frames are best bought ready-made.

236 Building a
Langstroth Beehive

244 Making a Warré
Beehive

245 Building a Top-Bar
Beehive

7 Building Beehives

Building a Langstroth Beehive

Langstroth Beehive, Exploded View

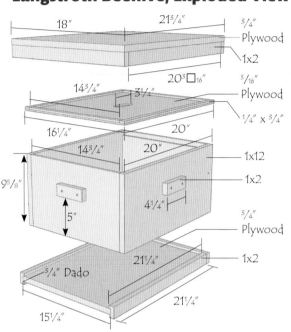

18"   21³⁄₄"   ³⁄₄"
Plywood
1x2
20³⁄₁₆"   ³⁄₁₆"
14³⁄₄"   3¼"   Plywood
¼" x ³⁄₄"
16¼"   20"
14³⁄₄"   20"   1x12
1x2
9⁵⁄₈"   5"   4¼"
³⁄₄"   Plywood
21¼"   1x2
³⁄₄" Dado
21¼"
15¼"

Tools

Circular saw with
   24-tooth carbide-tip
   blade
Measuring tape
Framing square or
   drywall T-square
Table saw
Saber saw
Putty knife or chisel
Hammer
Clamps
Drill-driver and bits
1¼-in. (32mm)
   hole-cutting saw
Sanding block

Materials

2 x 4-ft. (61 x 121.9cm)
   piece of ¾-in.
   (19mm) plywood
2 x 2-ft. (61 x 61cm)
   piece of ³⁄₁₆-in. (5mm)
   plywood
6-ft. (1.8m) 1x12 (straight
   and uncupped)
2 8-ft. (2.4m) 1x2s
Waterproof glue
5d (1¾-in. [44mm])
   galvanized box nails
³⁄₈-in. (10mm) brads
Ready-made

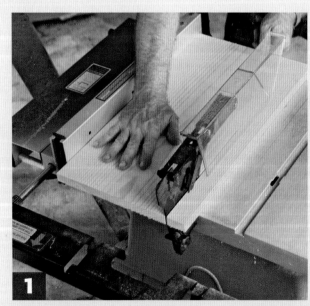

1

From a straight, uncupped 1x12, cut two pieces 20 in. (50.8cm) long and two pieces 14¾ in. (37.5cm) long for the sides of the super. Rip them to match the height 9⁵⁄₈ in. (24.4cm). of your frames. In this case, rip each piece to a width of 9⁵⁄₈ in. (24.4cm).

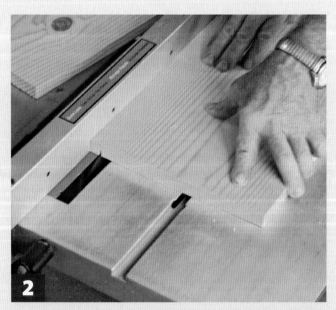

2

Using a table saw (shown) or router, cut a notch, called a rabbet, ⁵⁄₈ in. (15.9mm) deep and ³⁄₈ in. (9.5mm) wide into the 14¾-in. (37.5cm) sides to hold the frames. Set the blade of the saw so that it protrudes ³⁄₈ in. (9.5mm); set the fence for the ⁵⁄₈-in. (15.9mm) cut. Make a test cut on scrap; then make the cut in the actual side.

Complete the rabbet by making repeated cuts, moving the board away from the fence about ⅛ in. (3.2mm) for each cut. Remove as much of the wood as possible; when done, crack away the remaining slivers of wood. Then smooth the rabbet using a sander (inset).

Assemble the super using 5d (1¾-in. [44mm]) galvanized box nails to fasten its sides. Drill pilot holes first to avoid splitting the wood. Fasten a clamp at the bottom of the vertical sides to support them as you work. Apply waterproof glue, and nail the sides together.

It Makes Sense to Buy Frames

Frames for Langstroth beehives are immensely complicated and not worth making yourself unless you just plain love intricate woodworking. In addition, they need to be made of clear pine. Experienced beekeepers watch for online sales and stock up. Three sizes are available. The smaller the frame, the lighter the super.

Insert a frame to confirm that it slips easily into the super. It should hang from the rabbet and have a little play at each end. The super should accommodate 10 frames with a little "bee space" in between.

Continued

Building a Langstroth Beehive (cont'd)

6

Cut out the bottom board, the entry point for the bees. Cut a piece of ¾-in. (19mm) plywood to 15¼ x 21¼ in. (38.7 x 54cm). Use a framing square to lay it out. Cut the plywood using a circular saw equipped with a plywood blade or a 24-tooth carbide-tip blade.

7

Dado sidepieces using the same method as shown in Step 2 for the rabbets. Cut a ¾-in. (19mm)-wide dado centered on the sides of two 21¼-in. (54cm) pieces of 1x2. Use a putty knife or chisel to crack out the leftovers.

8

Fasten the bottom-board sides in place. Predrill, glue, and nail the dado sidepieces from Step 7 to the long sides of the bottom board. Hold a piece of 1x2 along one shorter end, and mark it for length. Cut it, and glue and nail it in place at one of the ends.

9

Make a frame for the outer cover by cutting an 18 x 21¾-in. (45.7 x 55.3cm) rectangle of ¾-in. (19mm) plywood. Using it as a guide, cut and nail together a lip of 1x2s for the cover. Drill pilot holes; then glue and nail the lip pieces together.

10

Attach the outer cover top by predrilling holes and starting 5d galvanized box nails into the top piece. Spread glue on the frame. Nail one edge; square up the top on the lip; and finish nailing.

11

Mark for cutting the inner cover from ³⁄₁₆-in. (5mm) plywood by using the completed super as a guide. Sawing on the outside of the line, cut out the piece.

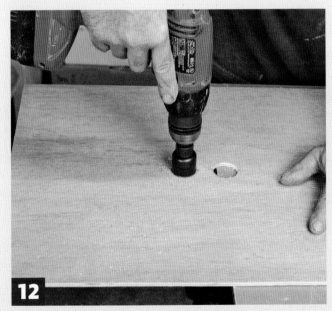

12

Make a mark centered on the long width of the piece. Next, mark for a centered lengthwise vent opening 3¼ in. (8.3cm) long. Using a 1¼-in. (32mm) hole saw, cut two holes so that their outer edges are at the ends of the 3¼-in.-long (8.3cm) line, centered on the line.

13

Complete the vent hole by marking lines joining the edges of the holes. Using a saber saw, complete the opening. Sand both sides of the hole (inset) so that it is rounded and smooth.

Continued

Building a Langstroth Beehive (cont'd)

14

Cut the inner cover edging out of 1-by ripped ¼ in. (6mm) wide and long enough for the four sides. (Scraps from ripping the 1x12s work for this.) Hold the longer pieces on the cover, mark them for trimming, cut them, add glue, and using ⅜-in. (10mm) brads, nail them in place.

15

Mark and cut the two shorter sides. Glue and nail them in place.

16

Flip the cover over, and nail from the other side to ensure that this fairly fragile component will hold together. Sand any rough edges, and test-fit it atop the super.

17

Assemble the components of your Langstroth beehive to make sure that everything fits. The outer cover should fit loosely with about ⅛-in. (3.2mm) clearance on all four sides.

18

Attach handles made from 1x2s cut 4–6 in. (10.2–15.2cm) long. Predrill holes for 1¼-in. (32mm) exterior screws. Add glue, and attach handles on all four sides, centered and about 3 in. (7.6cm) down from the top edge of the super.

19

Paint the exterior of the beehive only; do not paint any portion of the interior. Even nontoxic paint may be harmful to bees. The color choice is up to you. Some beekeepers camouflage their hives using dark greens. Others believe that bold blossom tones make bees feel welcome. Let the paint dry for a couple of weeks before using the hive.

Making a Warré Beehive

Named after its inventor, Abbé Emile Warré, the Warré beehive mimics the natural home bees choose, a hollow tree. Like the top-bar beehive (opposite page), the Warré uses bars instead of foundation-filled frames to let the bees hang their combs as they do in the wild. Like a tree trunk, a Warré hive is thin and tall. And like a tree trunk, it is designed to grow.

The bees are introduced to the top box, where they build combs to hang beard-like from the bars. The sawdust-filled "quilt" box absorbs moisture. As new bees hatch, the brooding colony moves downward. As boxes fill, you can add new ones so that honey is always rising, easily accessible to the beekeeper.

Backyard Homesteading (Creative Homeowner, 2012, ISBN: 978-1-58011-521-6) shows how to build one. Thorough Warré plans can also be found online. If you are not equipped to cut rabbets to hold the bars or indeed to rip any boards at all, however, consider the approach shown here. It doubles the sides to make the ledge for the bars. It also uses 1x2s for bars, just as the top-bar hive does. Use construction techniques similar to those shown for the Langstroth hive on pages 236-243.

Warré Beehive, Exploded View

This plan allows for 7 1x2 bars with ³/₈ in. (9.5mm) bee space between them and at each end. Use two ³/₈ in. (9.5mm) spacers to position each bar; then drill a ¹/₈ in. (3mm) hole and pound in a 4d finishing nail.

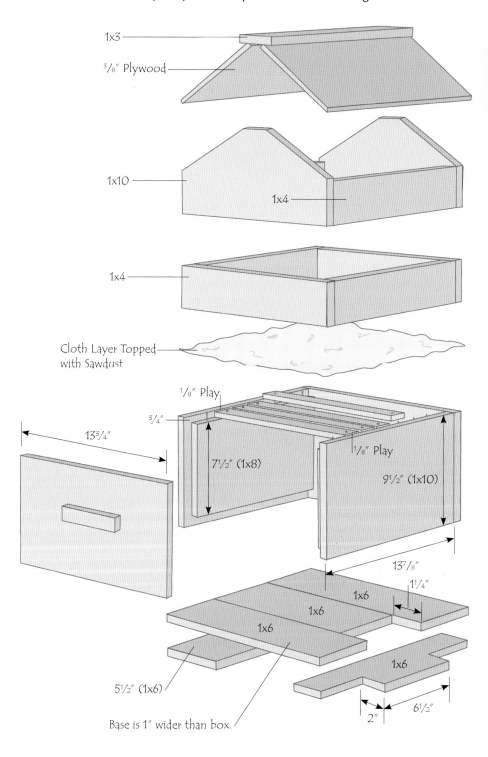

1x3
³/₈" Plywood
1x10
1x4
1x4
Cloth Layer Topped with Sawdust
¹/₈" Play
³/₄"
13³/₄"
7¹/₂" (1x8)
¹/₈" Play
9¹/₂" (1x10)
13⁷/₈"
1¹/₄"
1x6
1x6
1x6
1x6
5¹/₂" (1x6)
6¹/₂"
2"
Base is 1" wider than box.

Building a Top-Bar Beehive

Known also as the Kenyan or Tanzanian beehive, the top-bar hive is easy to work with and minimally disruptive to bees. To harvest honey, you need only remove the roof and lift out a single bar—no hefting of whole supers. This allows you to harvest throughout the season, a chance to enjoy different flavors of honey as new plants come into bloom.

It is also a flexible hive. The length can vary from a "nuc," or nucleus, hive holding only eight bars to a 42-bar hive roughly 66 in. (167.6cm) in length. The entrance can be at one end or centered in the hive, depending on your preference. The entrance itself can be a series 1⅛-in. (2.9cm) holes (corkable in the winter) or a slot ⅜ in. (9.5mm) high and 4–6 in. (10.2—15.2cm) wide.

The bars can be as simple as a 1x2 smeared with a bead of wax to encourage comb building to a slotted bar that accepts a strip of wax foundation—or anything in between.

Top-Bar Beehive, Exploded View

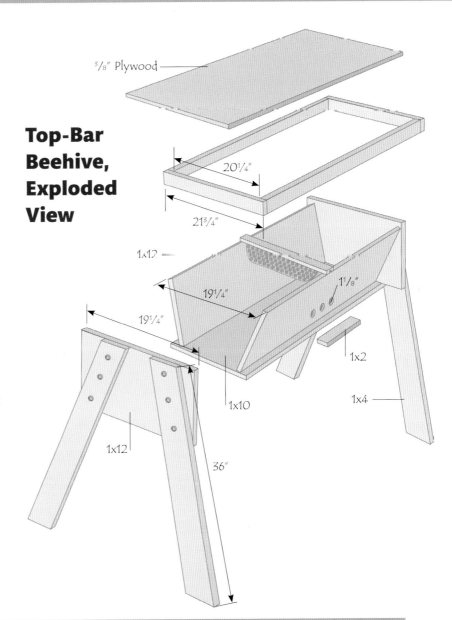

Figuring the Angles

Six angles of 120 degrees make up the cell of a honeycomb. By the slanting the sides of a top-bar beehive at this angle, you are encouraging bees to build a nicely rounded comb unattached to the sides of the hive.

Legs are optional—there is no harm in setting a top-bar hive on a stump or a couple of concrete blocks. Legs about 38 in. (96.5cm) long should be a comfortable working height for most beekeepers, but feel free to adjust them to a height that suits you.

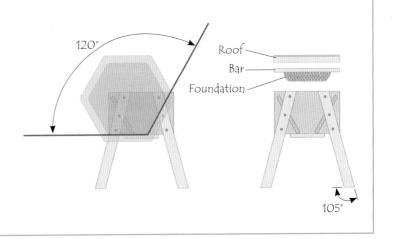

A HUMBLE GARDEN HOSE fed by an outdoor spigot is often enough to water the garden in the summer. You can even run a fairly extensive micro-irrigation system off a residential outdoor spigot—if you are careful to add a backflow preventer to it. Most such spigots are shut down and mothballed for the winter. But once you have livestock, you are going to need a spigot that operates year-round. It's inevitable: as your backyard homestead grows, you are going to need to upgrade your outdoor plumbing.

Electricity poses the same problem. Despite the ubiquity of powerful cordless tools like drill-drivers, some tools, such as full-size circular saws and electric chain saws, pull enough juice to require an extension cord. Prone to tangles, dangerous in wet weather, and always in just the right spot to trip you up, an extension cord is at best a short-term solution. Eventually you will want to add additional exterior receptacles, yard lights, heat for water dispensers, and supplemental lighting for poultry. That calls for safe, permanent installations. The need is even greater once you build a good-size shed, which calls for interior lighting and plenty of receptacles.

This chapter provides you with the essentials for getting water and power where you want it. As with all permanent improvements, be sure to check local code requirements before embarking on these projects.

Spigots and Hydrants

The outdoor spigot, variously known as a hose bibb, sillcock, or bibcock, is a valve with external hose threads on its spout just right for attaching a hose. Most such spigots are brass and use a compression mechanism for controlling water. For all-season, heavy-duty use, the freeze-proof yard hydrant (pictured on the opposite page), has a subterranean valve and is ideal for watering livestock. See pages 252-253 for how to install it.

Male threads on some spigots allow you to attach them to elbows and other fittings, handy if you need a stand-alone water source.

Female threads on other spigots allow you to fasten the faucet directly to supply pipes—often from the sill of the house, thus sillcock.

rd Homesteaders

250 No-Sweat Sweating

251 Water Where You Need It

252 Installing an Anti-Freeze Spigot

254 Repairing a Freeze-Proof Spigot

255 Adding a Freeze-Proof Yard Hydrant

257 Your Friend, the GFCI Receptacle

259 Installing a GFCI Receptacle

260 Running Outdoor Conduit and Cable

261 Outdoor Electrical Boxes

262 Installing Outdoor Receptacles and Switches

263 Conduit and Fittings

264 Installing UF Cable

266 Installing a Stand-Alone Receptacle

267 Adding Supplemental Light for Poultry

No-Sweat Sweating

Cut the copper tube to length using a tubing cutter. You can also use a hacksaw, but you'll have to clean up any burrs.

Brighten the tubing using a wire brush or sandpaper. Burnish the inside of the fitting as well (inset).

Tools

Hacksaw or tubing
  cutter
Propane torch
Wire brush or
  sandpaper
Flux brush
Work gloves
Eye protection

Materials

Copper tube
Fittings
Solder
Flux
Rags

While joining steel pipe is a simple matter of wrapping joint sealing tape clockwise around the threads and attaching a fitting with a wrench, connecting copper tubing requires a soldering technique known as "sweating." Keys to the technique are clean, dry tubing and fittings, plus knowing when the joint is hot enough for the solder to be drawn in by the flux. Once you get the knack, you will find that sweating is pure magic.

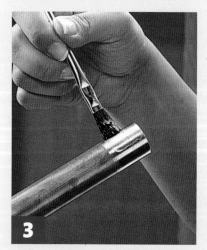

Apply flux to mating surfaces so that the molten solder will be drawn completely into the joint.

Assemble the connection, and apply heat evenly to the entire joint. Wear gloves, or use clamps to handle heated tubing.

Touch solder to the joint when the copper darkens slightly; the solder will melt and fill the joint. Remove excess solder using a rag.

Water Where You Need It

Hauling water to your chickens or goats loses its romance quickly, especially when you have splashed your pant leg and shoes a few times. Sooner or later you are going to want to pipe the water to where you need it. That is going to require changes to your plumbing system.

First, you need to be sure that no water contaminated with farmyard waste will be siphoned into your household plumbing. How does this happen? Remember that the water in your house is under pressure. If that pressure is interrupted—by the pull of a power washer, the failure of the well pump, and so on—suction results. If you have a hose lying in the goat pen or a micro-irrigation system running from an outdoor spigot to your compost-laden garden, that suction can introduce pathogens into your household water. This phenomenon is called backflow, or back siphonage. You will need to install a backflow preventer or vacuum breaker to any outdoor spigot that you install. (See photo at right.)

Second, you need to protect the line from freezing. A two-valve assembly (bottom left) is often used for an outdoor spigot on an exterior wall, with a second, in-line valve just inside the house. When it begins to get cold, you shut off the inside valve, drain the outside valve, and let winter do its worst. This setup has a disadvantage: you have to remember to drain the assembly before the first frost. An extended-valve setup (bottom right), or freeze-proof spigot, solves this problem. The valve is set back well into the house, defended from freezing. When installing a freeze-proof spigot, prop up the rear of the chamber so that water can drain through the spout. Freeze-proof spigots come in lengths from 6 to 36 inches. Choose one that places the valve a foot or so inside the house.

You can attach an aftermarket vacuum breaker to a spigot to prevent backflow into the house.

Freeze-Proof Outdoor Spigot Setups

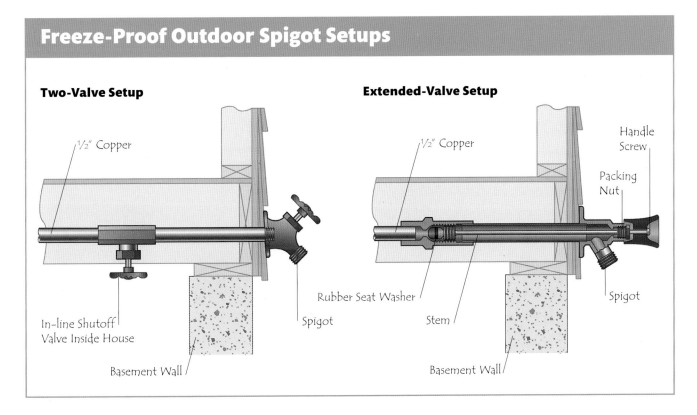

Two-Valve Setup

- 1/2" Copper
- In-line Shutoff Valve Inside House
- Basement Wall
- Spigot

Extended-Valve Setup

- 1/2" Copper
- Handle Screw
- Packing Nut
- Rubber Seat Washer
- Stem
- Basement Wall
- Spigot

Installing an Anti-Freeze Spigot

To protect an outside spigot against freezing, replace it with a special anti-freeze valve. You can turn the water on and off outside, but the long stem of the valve extends through the wall and controls a valve inside the house. Water doesn't stand in the portion of the pipe or spigot outside the wall where it could freeze. See page 253 for how to solder copper joints. Once you have installed the spigot, caulk and insulate the hole through the sill framing.

Tools	Materials
Propane torch	Anti-freeze spigot
Wire brush	Screws
Caulking gun	Solder and flux
Work gloves	Braces
Eye protection	Caulk
Tubing cutter	Insulation
Reciprocating saw (optional)	

Freeze Prevention

To protect pipes from freezing, insulate the pipes and seal air leaks near plumbing using caulk and spray-foam insulation. If you believe that your pipes might still be at risk, install a heating cable, which looks like an extension cord. Follow the manufacturer's installation instructions. Some cables need only be mounted alongside the pipe; other cables must be wrapped around the pipe. If you wrap the pipe, don't overlap the cable. The wires can overheat and melt through the cable insulation.

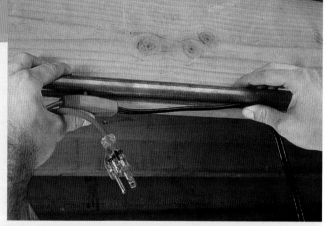

New heating cables have a built-in thermostat that can be left plugged in. Never wrap a cable over itself.

Freeze Repair

Sadly, too often you don't know a pipe is prone to freezing until you get hit with a severe cold snap and it ruptures. If you cannot replace the water pipe right away, make this temporary repair. Begin by turning off the stop valve if there is one or by shutting off the main valve for the house. If you have a well with a pressure tank, remember that water will continue to flow even if you shut off the well pump.

File off any ragged edges. Wrap the pipe with a piece of thick rubber sheet or a section of garden hose that you have slit.

Attach banded clamps over the patch at each end of the rupture (and in the middle if it is a bad break), and tighten them.

Installing an Anti-Freeze Spigot

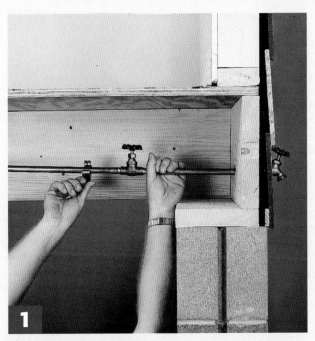

1

Shut off the water supply, and drain the system. Cut out the old spigot piping, including the shutoff valve, using a tubing cutter.

2

Enlarge the wall opening if needed, and insert the freeze-proof spigot through the siding and band joist. Secure it using screws.

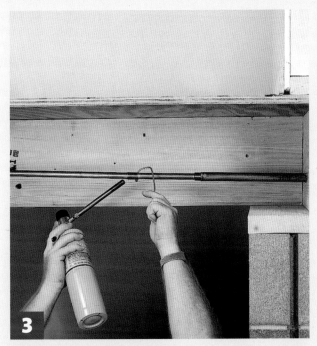

3

Sweat the copper supply line to the spigot using flux and lead-free solder. Brace the line to ensure adequate outward drainage.

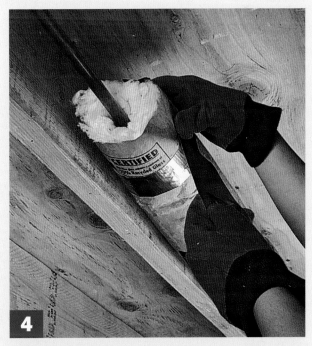

4

Insulate the line using split foam tubing or batts of insulation. Use caulk or spray-foam insulation to fill the hole in the joist.

Repairing a Freeze-Proof Spigot

Tools	Materials
Screwdriver Adjustable wrench Shop vac	Replacement washer

Repairing a freeze-proof spigot is not much more difficult than repairing a standard valve-stem-and-seat faucet. The only real difference is that any leftover rubber-washer pieces that may have broken off the end of the valve stem are harder to fish out of the faucet body. If you leave debris like this inside the faucet, the new replacement washer will not seat properly. Turn on the water briefly to flush debris. Another option is to duct-tape the end of a wet/dry shop-vac hose to the end of the spigot, and vacuum out any loose material.

1

Begin the repair by removing the handle and loosening the bonnet nut using an adjustable wrench. Look for an arrow that indicates the proper turning direction, either clockwise or counterclockwise. Once this nut is loose, pull out the valve stem.

2

Inspect the end of the valve stem to check whether the old washer is still in place or not. If it is, there should not be any washer debris in the body of the spigot. Replace the worn washer with a new one of the same size.

3

Slide the handle onto the end of the valve stem, but do not attach it. The handle makes stem-into-body threading easier. When the valve is in, remove the handle; tighten the bonnet nut; and reinstall the handle. Test the valve.

Adding a Freeze-Proof Yard Hydrant

The most convenient water source for a backyard homestead is a freeze-proof yard hydrant. Like a freeze-proof sillcock, a hydrant has a long stem and a drain chamber. Instead of draining unneeded water through the spout, a hydrant drains downward into a drainage pit near the shutoff mechanism. When you raise the on-off lever, it lifts the long stem and the rubber stopper attached to it. This allows water to fill the riser and flow from the spout. When you lower the lever, the stem forces the stopper back into its seat, blocking the flow. The water left in the riser then drains through a small opening just above the stop mechanism; this keeps the riser from freezing. Gravel eases the excess water into the soil. Because you bury the stop mechanism and drain fitting below frost level—typically 3 to 6 ft. (91.4cm x 1.8m) deep—you can use the hydrant year-round.

To repair a freeze-proof yard hydrant, shut off the water supply. Thread the handle-and-spout assembly counterclockwise, and lift the assembly and stem from the riser. Remove the worn stopper from the stem, and install a new one. When you thread the handle-and-spout assembly off or on, be sure to stabilize the riser using a second wrench. This keeps you from undoing the piping connection at the bottom of the trench.

Find the Frost Level

Your local building department can provide the frost depth in your area so that you can set your hydrant drain opening and gravel reservoir out of reach of freezing ground temperatures. Once you dig even a few feet beyond the frost level, the earth's temperature remains relatively mild—ranging from the high 30s (Fahrenheit) in the northern United States up to the mid 70s in the extreme south.

Freeze-Proof Yard Hydrant Elevation

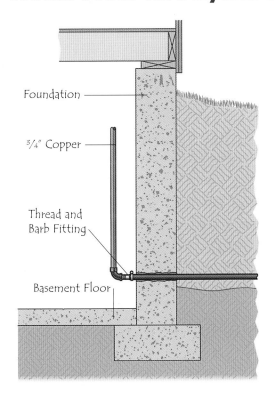

Foundation

¾" Copper

Thread and Barb Fitting

Basement Floor

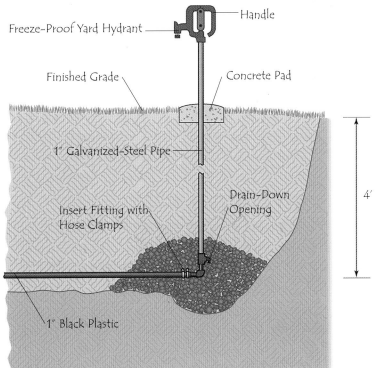

Freeze-Proof Yard Hydrant

Handle

Finished Grade

Concrete Pad

1" Galvanized-Steel Pipe

Insert Fitting with Hose Clamps

Drain-Down Opening

4'

1" Black Plastic

Tunneling under Obstacles

Your yard, like most, may have a few obstacles, including sidewalks and tree roots, under which you may need to tunnel. With some obstructions you can dig a hole on each side and tunnel directly through using a tile spade. For a longer reach, dig a hole on each side and burrow across using a sluice pipe: Just attach a pointed spray nozzle to one end of a PVC pipe and a hose fitting to the other. Connect the hose, and turn on the water. Push the pipe a few inches into the soil and then pull back. Continue pushing and pulling to tunnel to the other side. The sluice water will carry the soil toward you as the pipe progresses across the span.

Make a sluice pipe for boring under sidewalks and other obstructions. The brass nozzle creates enough force to cut into the soil and bear the waste back toward you.

Solving the Drain Problem

If you have a shed and want to use a utility sink occasionally, you may be able to install a dry well instead of a standard drainpipe and sewage line. This simple solution holds gray water until it can percolate into the soil. To make a dry well, buy a 10-gallon plastic bucket and perforate it with 1-in. (2.5cm) holes roughly 2 in. (5.1cm) apart. Near its upper lip, bore a hole for a PVC drainpipe. Dig a hole about 9 in. (22.9cm) deeper than the height of the bucket. Add 3 in. (7.6cm) of gravel to the bottom of the hole; set the bucket in place; and install the drainpipe. Fill the bucket with stones, and cover it with landscaping fabric. Fill the remaining 6-in. (15.2cm) hole with soil, and replace the sod.

Your Friend, the GFCI Receptacle

Any receptacle you add to a coop or shed should be protected by a ground-fault circuit interrupter (GFCI). In fact, GFCI protection is required by the National Electrical Code (NEC) for any damp environment.

A GFCI (built into either a circuit breaker or receptacle) prevents electrocution caused by an accident or equipment malfunction. In a general-purpose 120-volt household circuit, current moves along two insulated wires—one black and one white. The black wire delivers power to the device or appliance; the white wire returns it to the electrical panel. As long as these two current flows remain equal, the circuit operates normally and safely. If a portion of the return current is missing, or "faulted," however, a GFCI will de-energize the circuit in 1/25th to 1/30th of a second—25 to 30 times faster than a heartbeat. In this fraction of a second, you may receive a jolt but not the potentially lethal shock that would occur without the protection of a GFCI.

No Ground Fault

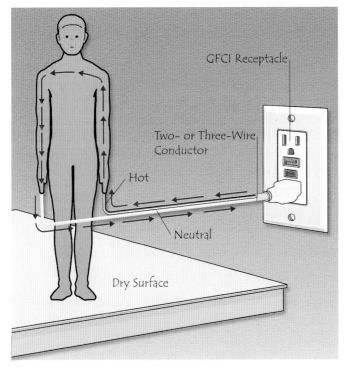

Ground Fault

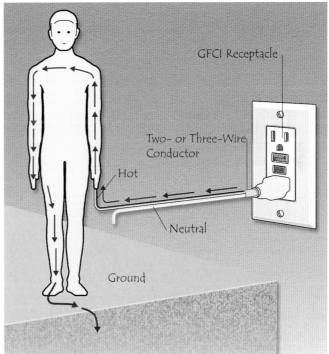

If an electrical current flows through your body from a hot wire to a neutral wire, this completes an electrical circuit—just as though you were a power tool or fixture. In this case, a ground-fault circuit interrupter cannot save you from being electrocuted because it cannot distinguish you from, say, your circular saw. If you hold only one wire, however, the resulting imbalance in current entering and leaving the circuit will trip the GFCI and protect you from serious shock or electrocution.

A GFCI receptacle is not foolproof, however. For a GFCI to succeed, a ground-fault must first occur. This happens when current flows out of the normal circuit to a ground pathway, causing the imbalance between the black and white wires mentioned earlier. In this instance, if you hold the black and white wires and you are not grounded, the GFCI will not function properly because it has no way of distinguishing your body from any other current-drawing device. A breaker or fuse is only designed to protect your household wiring against excessive current—it is not designed to protect you.

A GFCI receptacle resembles a conventional receptacle, except that it has RESET and TEST buttons. A GFCI can also be installed at the panel box as a circuit breaker. This kind of GFCI has a test button; when tripped, the switch flips only halfway off to break the circuit. To reset the circuit, you must switch the breaker completely off and then flip it back on again. A GFCI receptacle is less expensive than a breaker-type GFCI and has the advantage of letting you reset a circuit at the point of use.

A GFCI receptacle has both TEST and RESET buttons. When a ground fault occurs or you perform a test, the RESET button will pop out. Once a fault is eliminated or the test is completed, press the button back in to reset the circuit.

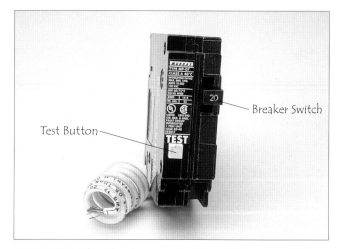

A GFCI circuit breaker has a TEST button, but no RESET button. To reset a GFCI breaker, first push the switch to the OFF position; then flip it back to the ON position.

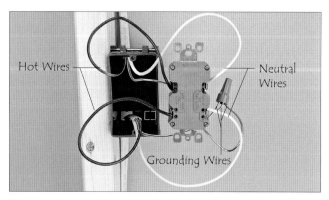

If you want several receptacles farther down the circuit, or downstream, from the GFCI receptacle to also have GFCI protection, you may decide to use this method of wiring for multiple locations: connect hot and neutral wires from the breaker box to the LINE terminals; connect hot and neutral wires carrying power to other devices to the LOAD terminals.

To install the circuit-breaker type of GFCI, right, simply insert the device into the panel box in the same way as a conventional circuit breaker. Then connect the black and white load wires from the circuit you wish to protect. Connect the white corkscrew wire attached to the GFCI circuit breaker to the white neutral bus in the panel. Some panels have separate ground bus bars.

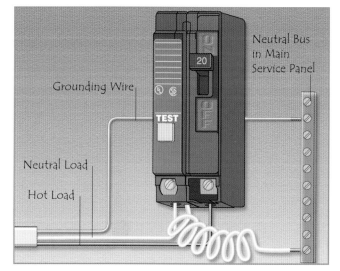

Installing a GFCI Receptacle

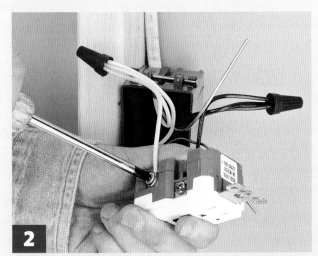

1

Pull the power cable and the cable leading to other receptacles into the box. Strip the sheathing from the cable and the insulation from the ends of all of the wires. Join the black cable wires with a black pigtail using a wire connector. Then attach the black pigtail to the brass (or dark) LINE terminal screw on the GFCI receptacle.

Tools

Insulated screwdriver
Diagonal cutting pliers
Needle-nose pliers
Wire stripper
Neon circuit tester
Cable ripper

Materials

GFCI receptacle
Receptacle box
12/2G NM cable
Wire connectors
Copper grounding wire
Grounding pigtail and
    screw (for metal boxes)
Cable clamps (for metal
    boxes)

Installing a GFCI receptacle isn't much harder than installing a standard receptacle. The only difference is that the GFCI unit has two sets of terminal screws, each with a different purpose. One set is marked LINE and is used for the incoming power line; the other is marked LOAD and may be used for the outgoing wires that connect to downstream receptacles in the circuit. Before doing this project, test that the circuit is on; shut the circuit off at the breaker box; then test again to confirm that it is off.

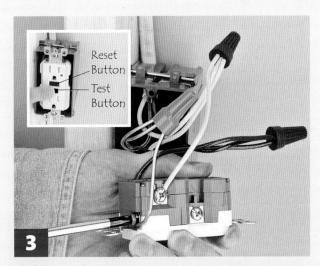

2

Join the white wires from both cables with a white pigtail, and attach the pigtail to the silver (or white) LINE screw on the receptacle. NOTE: downstream receptacles will not be GFCI-protected in this scenario; you will have to install a GFCI receptacle at each location, generally considered better pratice than using the LOAD terminal.

Reset
Button
Test
Button

3

Finish up the wire connections by joining the grounding wires from the cables with a grounding pigtail using a wire connector. Then attach the pigtail to the receptacle's grounding screw. Install the receptacle, and test the RESET button by pushing the TEST button. The RESET button should pop out.

Running Outdoor Conduit and Cable

Cable strung overhead from poles is unsightly, comes down in storms, and is harder to install than you might think. Underground cable or conduit is out of sight and worry free. Once you have dug the trench, you can pull underground cable through protective conduit or, if the cable is rated for such use, bury the cable without conduit.

Underground feeder and branch-circuit cable, known as UF cable, is designated for outdoor wiring because it is weatherproof and suitable for direct burial. UF cable looks somewhat like ordinary NM cable, so look for the UF designation, which is printed on the sheathing. The cable wires are molded into plastic rather than wrapped in paper and then sheathed in plastic as are NM cable wires. For that reason, UF cable can be challenging to strip. Any UF cable run aboveground must be protected by conduit.

You must bury direct-burial cable deeply enough that it is protected from routine digging, yet not so deeply that trenching may interfere with existing water or power lines. The National Electrical Code (NEC) specifies minimum depth requirements for underground cable: 24 in. (61cm) for direct-burial cable, 18 inches for rigid nonmetallic conduit, and 6 in. (15.2cm) for rigid and intermediate metal conduit (IMC). (See the table opposite.) If a GFCI circuit breaker powers the cable, you may be allowed to trench less deeply, but it is usually worth the extra trouble to get it low enough that you won't later accidentally cut into it.

ENT (Electrical Nonmetallic Tubing)

Flexible nonmetallic conduit offers limited protection for underground cable.

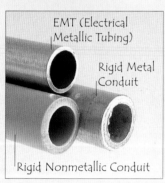
EMT (Electrical Metallic Tubing)
Rigid Metal Conduit
Rigid Nonmetallic Conduit

Rigid conduit affords extra protection for underground cable, but metal types can eventually fall prey to water penetration and corrosion. The cement-welded fittings of nonmetallic conduit offer better protection from moisture.

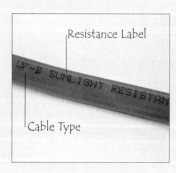
Resistance Label
Cable Type

Type UF (underground feeder) cable is designed for direct burial underground. The sheathing label indicates whether it is also sunlight and corrosion resistant.

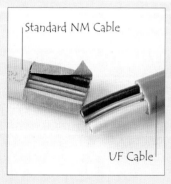
Standard NM Cable
UF Cable

UF cable doesn't have paper insulation between the wires and outer sheathing. A thermoplastic coating encases the wires, making them water resistant but difficult to strip.

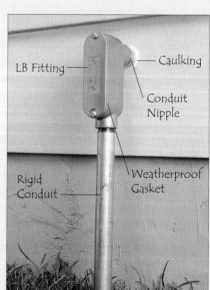
LB Fitting
Caulking
Conduit Nipple
Weatherproof Gasket
Rigid Conduit

Outdoor cable run underground must be protected in rigid conduit where it enters or emerges from the trench.

Underground Cable-Depth Requirements

Condition	Direct-Burial Cable	Rigid Nonmetallic Conduit (PVC)	Rigid and IMC Conduit	
In open soil—pedestrian traffic only	24" (61cm)	18" (45.7cm)	6" (15.2cm)	Code requires that there be a minimum distance between the topmost surface of an underground cable or conduit and the top surface of the finished grade or other cover above the cable or conduit.
In trenches below 2" (5.1cm) of concrete	18" (45.7cm)	12" (30.5cm)	6" (15.2cm)	
Under streets, highways, roads, alleys, driveways, and parking lots	24" (61cm)	24" (61cm)	24" (61cm)	
1- & 2-family dwelling driveways and outdoor areas; used for dwelling related purposes only	18" (45.7cm)	18" (45.7cm)	18" (45.7cm)	

Outdoor Electrical Boxes

Outdoor electrical boxes are either rain-tight or watertight. Rain-tight boxes typically have spring-loaded, self-closing covers, but they are not waterproof. This kind of box has a gasket seal and is rated for wet locations as long as the cover is kept closed. Mount a rain-tight box out of the way of driving rains or flooding. Watertight boxes, on the other hand, are sealed with a waterproof gasket and can withstand a soaking rain or temporary saturation. These boxes are rated for wet locations.

The Weatherproof Difference

Outdoor electrical materials and equipment, such as fixtures, electrical boxes, receptacles, connectors, and fittings, must be manufactured not only to meet NEC requirements but also to resist the elements. Outdoor electrical equipment must be weatherproof and, in some cases, watertight. That is why you use different materials and equipment for outdoor electrical work than for indoor work.

A rain-tight box is rated for wet locations only if the cover remains closed. It is not fully waterproof.

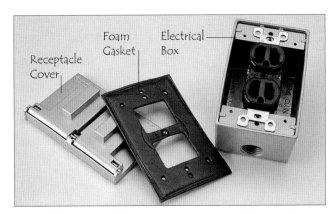

Watertight receptacle boxes use waterproof foam gaskets to create an impermeable seal. Receptacle covers snap shut. Switch covers have watertight levers.

259

Installing Outdoor Receptacles and Switches

Any receptacles that provide outdoor power, even if they are in a shed or other outbuilding, must have ground-fault-circuit-interrupter (GFCI) protection. (See pages 257-259.) Although you may use GFCI receptacles, they tend to trip intermittently when exposed to the weather. It is better to have your outdoor branch circuit powered by a cable connected directly to a GFCI circuit breaker.

Every residence must have at least one receptacle installed at the front and back of the house. These receptacles must be within 6½ ft. (2m) of the finished grade. In addition, any outdoor receptacle in unattended use, such as one that supplies power to a pump motor, must have a weatherproof box and a cover that protects the box even when the plug is in the receptacle. Receptacle covers are available for both vertical and horizontal installations and are either on the device in the box or attached to the box itself.

You may mount an outdoor receptacle on a wall, post, or any secure location. If you plan to install a weatherproof outdoor receptacle in the yard or other open area, you may attach it to a pressure-treated wooden post or mount it on the ends of two ½-inch-diameter

(1.3cm) sections of galvanized rigid metal conduit threaded on one end and anchored in concrete at the other (photo below and illustration opposite).

Weatherproof boxes and covers are also required to protect outdoor switches from exposure to the elements. Covers to single-, double-, and triple-gang boxes operated by toggle levers are available for outdoor switches, and there is also a cover for a combination single-pole switch with a duplex receptacle.

A freestanding receptacle box supported by rigid metal conduit must be mounted at least 12 in. (30.5cm) but no more than 18 in. (45.7cm) aboveground. It should have secondary support, such as a second conduit.

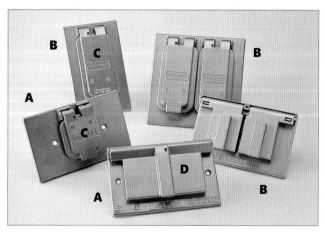

Receptacle cover plates for vertical or horizontal boxes may be box-mounted **(A)** or device-mounted **(B).** Cover types can be snap-shut **(C),** screw cap, or flip-top **(D).**

Weatherproof boxes and cover plates are available for single-pole, double-pole, and three-way switches. Covers also exist for switch/pilot lights, such as **(A)** double gang cover plates, **(B)** switch/receptacle cover plates, or **(C)** single-gang cover plates.

The Well-Mounted Receptacle

Centerline

12″ Minimum
18″ Maximum

Rigid Metal Conduit

Secondary Support

Concrete Anchor

24″ Minimum

Conduit Sweep

UF Cable with Expansion Loop

A concrete-anchored, rigid metal conduit may be used as a support on which to mount a weatherproof outdoor receptacle (with secondary support). Check local code for above-grade minimum and maximum height requirements.

Wet-Rated Weatherproof Boxes

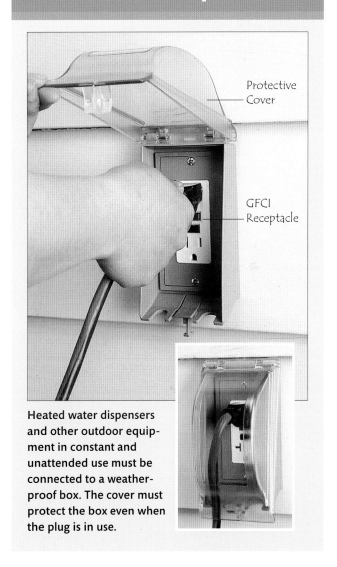

Protective Cover

GFCI Receptacle

Heated water dispensers and other outdoor equipment in constant and unattended use must be connected to a weatherproof box. The cover must protect the box even when the plug is in use.

Conduit and Fittings

Outdoor wiring is typically protected by rigid conduit aboveground and wherever it enters or emerges from underground trenching. Rigid and intermediate metallic conduit (IMC) are most commonly used, but most local codes permit the use of rigid nonmetallic conduit, which is made of Schedule 80 polyvinyl chloride (PVC). Regardless of which type of rigid conduit you use, you will need special connectors: bushings for straight pieces and elbow connections, locknuts, offsets, and various couplings. Be sure that the connectors you select match the material and category of conduit you are using.

At the point where cable runs through the exterior wall of a shed or other structure, you will need a special L-shaped connector called an LB conduit body. An LB encloses the joint between your indoor cable and the outdoor UF cable running down the side of your house and into an underground trench. LB units contain a gasket that seals the cable connection against the weather.

Another kind of fitting that you may find useful is a box extension, or extender, which is used to increase the volume of an existing outdoor receptacle or junction box when you must tap into it to bring power where you need it. Using a box extension makes it possible for you to avoid extensive rewiring or renovation work.

Trench Wisely

Don't dig a trench without first knowing what is underground. If you excavate at random, you may unwittingly cut into a sewer or water pipe or a telephone, cable TC, or electrical power line. Before you do any digging, check with your local utility company and have it mark the location of all underground utility lines near where you plan to dig. In most areas, you are required by law to inform your utility company and secure its approval before you excavate. Once the utility company clears you to excavate, you can dig your trench using a shovel, mattock, backhoe, trencher, or any other suitable equipment. When you run UF cable in a trench, be sure to leave a slack loop for expansion wherever the cable enters or leaves the conduit. For depth requirements, see the table on page 261.

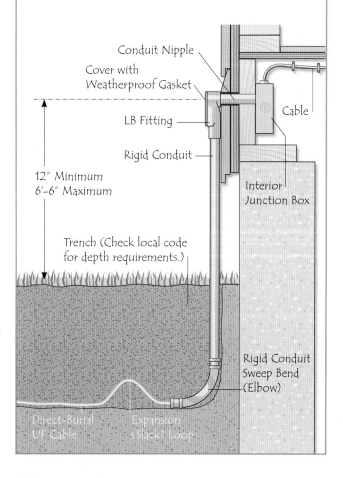

Conduit Nipple

Cover with Weatherproof Gasket

LB Fitting

Rigid Conduit

Cable

Interior Junction Box

12" Minimum
6'-6" Maximum

Trench (Check local code for depth requirements.)

Rigid Conduit Sweep Bend (Elbow)

Direct-Burial UF Cable

Expansion (Slack) Loop

Installing UF Cable

1

Lay out the your trench from the LB fitting on the house to the cable's destination point, using wooden stakes and mason's string. Dig the trench, setting aside the sod so that you can put it back in place after refilling the trench.

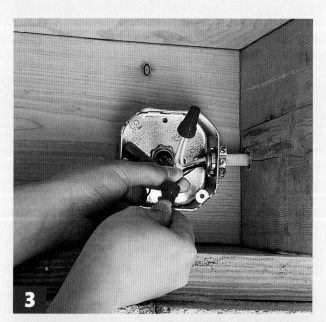

3

Attach the NM power cable from the breaker panel to the UF cable. Connect the white wires, black wires, and grounding wires (to a pigtail) using separate wire connectors. Attach the free end of the grounding pigtail to the grounding screw inside the junction box.

Tools	Materials
Round-head shovel	6-mil plastic sheeting
Adjustable pliers	Conduit compression
Baby sledge	connectors and
Chalk-line box	bushings
Mason's string	Stakes
Work gloves	UF direct-burial cable
	Wire connectors
	LB fitting
	Rigid conduit

UF cable is waterproof and specifically designed to be installed underground. The plastic insulation that encases the wires is rugged, but it still needs some protection when it is exposed to possible damage. That is why the code calls for rigid conduit protection above grade and where the cable enters or leaves a building.

2

Attach a conduit sweep to the conduit that's connected to the LB fitting. Install a plastic bushing on the open end of the sweep to protect the cable from abrasion. Feed the UF cable into the sweep, up through the conduit and LB fitting, through the house wall, and into the junction box.

4

Continue laying the UF cable, being sure to form an expansion loop in the cable next to the conduit sweep. Add loops anywhere the cable enters or leaves rigid conduit.

5

When you have finished the cable installation, carefully refill the trench; tamp it firmly; and replace the sod. Tamp the sod as well, and water it.

263

Installing a Stand-Alone Receptacle

Tools	Materials
Insulated screwdrivers	UF direct-burial cable
Multipurpose tool	Receptacle (GFCI if no
Needle-nose pliers	GFCI breaker)
Utility knife	Weatherproof box
Adjustable pliers	Rigid conduit
	Conduit compression
	connectors (if metal)
	Mounting ears
	Conduit sweep

As you now know, outdoor receptacles must be protected by a GFCI. But a GFCI receptacle may trip unexpectedly due to moisture buildup, so it may be worth your while protect the circuit with a GFCI breaker and install a standard receptacle. Support the box with steel conduit, or attach it to a post. Mount the receptacle box no lower than 12 in. (30.5cm) and no higher than 18 in. (45.7cm) above grade. Test that the circuit is on; shut off the circuit at the breaker box; then test again to confirm that it is off.

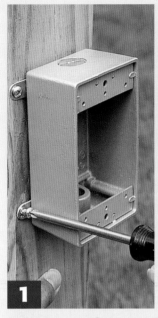

1

Mount an outdoor receptacle on a post or the side of a shed. Screw the mounting ears securely (left). Or attach the receptacle to rigid metal conduit anchored in concrete (right).

2

Install a plastic bushing in the conduit to prevent the cable from chafing against sharp edges. Pull the cable so that about 8 in. (20.3cm) extends from the box. Strip the thermoplastic coating by cutting the plastic with a utility knife, being careful not to cut the wire insulation (inset). Pull the plastic off, and strip ¾ in. (1.9cm) of insulation from each wire.

3

Make the connections by first attaching the grounding wire to the grounding screw on the side of the receptacle. Then, attach the black wire to the brass screw terminal and the white wire to the silver screw terminal. Fold the wires into the box and screw the receptacle in place. Finally, position the foam gasket and attach the box cover (inset).

Adding Supplemental Light for Poultry

Tools	Materials
Cordless drill-driver and bits	Exterior fixture
Phillips screwdriver	40-watt or greater bulb
Small standard screwdriver	Wire connectors
Stripping pliers	2–3 ft. (61–91.4cm) of heavy-duty electrical cord
	Plug, timer
	Mounting screws
	Extension cord plugged into GFCI receptacle

A light controlled by a timer counteracts the drop in egg production due to short winter days. Adding a baffle made of a scrap of plywood decreases glare.

Chickens produce more eggs during the short days of winter if they are exposed to supplemental light. The exterior fixture shown in this project comes with a globe that completely covers the bulb. The setup should be inside the coop or otherwise sheltered, but the potential humidity, dust, and general mess calls for exterior-grade supplies.

This project assumes that you don't have power running to your hen house but will instead use an extension cord plugged into a GFCI receptacle.

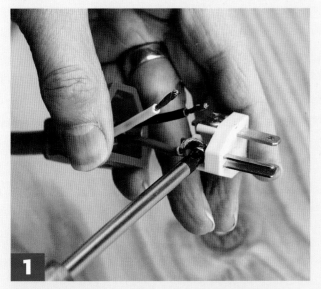

1

Connect a plug to a piece of heavy-duty electrical cord long enough to reach from the fixture to the timer that you will use. Use a utility-grade plug that has a grounding prong. You will attach the plug to the timer.

2

Mount the fixture box on the interior of your coop in a spot that is away from the hens. Choose a location so that the electrical cord can easily reach the timer. Bore a hole for the cord.

Continued

Adding Supplemental Light for Poultry (cont'd)

3

Insert the heavy-duty electrical cord into the fixture box, and remove 6 in. (15.2cm) of sheathing. Attach the green wire to the fixture's copper grounding wire. Then connect the cord's white wire to the fixture's white wire and black to black using wire connectors.

4

Assemble the fixture, and install the bulb and globe. Use a 25-W bulb—greater wattage for larger coops. An incandescent bulb uses more electricity but also provides more heat. If you use a compact fluorescent, pick one rated for outdoors so it will not dim in cold weather.

5

Install the timer in a sheltered spot that is easy for you to get to. This holiday timer is inexpensive and designed for outdoor use. Unlike some low-cost timers, it can be set for a few hours in the morning and a few in the evening, just what your hens need.

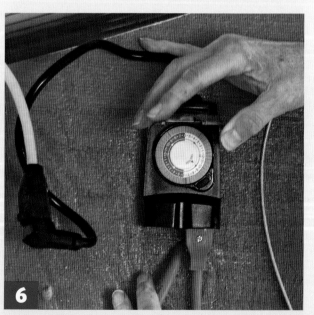

6

Set the timer to supplement both morning and evening light, and plug it in. Ensure that there will be light roughly equal to the longest day in your region but not more than 16 hrs. Work up to the desired duration by 15-min. increments every week.

Resources

American Dairy Goat Association

www.adga.org

Though aimed at commercial producers, this site is loaded with useful information about raising goats for milk. The page "Care and Management of Dairy Goats" offers useful information on housing and pasture requirements.

The Aquaponic Source Inc.

www.theaquaponicsource.com

The "Learn about Aquaponics" page alone is worth visiting this site. You will also find many hard-to-source products needed for setting up your system.

Backyard Chickens

www.backyardchickens.com

This site should be your first stop for brainstorming coop configurations and styles. You will find scores of coops, often with step-by-step photos of construction—sometimes even plans. You will also find a lively forum for all of your chicken-keeping questions.

Backyard Hive

www.backyardhive.com

You will find plenty of general beekeeping information at this site, with an emphasis on the top-bar beehive. The site has a mission: Empower backyard beekeepers to help bring back the feral bee population and improve the genetic diversity of honeybees.

Bergey Windpower Co.

www.bergey.com

In business since 1977, Bergey has its roots in aviation—an ideal background for engineering efficient, long-lasting turbines. Check out its Wind School for detailed background on homestead-scale turbine theory and practice, including "90 Second Expert" introductory articles.

Drill Your Own Well

drillyourownwell.com

If you are considering drilling a new well, check this site for a wealth of information on how to do it yourself. It is an education in itself on well drilling, pumping, and general water source troubleshooting.

FarmTek

www.farmtek.com/farm/supplies/home

You will find just about everything you need—and a lot more—from this online farm supply house. Expect to find innovative items like solar-powered fans, tension-fabric buildings, mini-greenhouses, hydroponic supplies, recycled-plastic raised beds, and farming solutions that you didn't even know existed.

Gempler's

www.gemplers.com/agriculture-supplies

As your backyard homestead grows to mini-farm scale, you may need to upgrade your watering, feeding, and even insect-control equipment. The site also offers small tractor accessories and a variety of fencing supplies.

Hobby Farms

www.hobbyfarms.com

From *Hobby Farms* magazine, this entertaining and informative site offers a range of topics relevant to an aspiring backyard homesteader. The livestock breed profiles are a handy start in choosing what animals might be right for you.

Irrigation Tutorials

www.irrigationtutorials.com

Visit this site for detailed tutorials on planning and installing drip irrigation systems, including reviews of various product types.

Missouri Wind and Solar

windandsolar.com

Covering large- and small-scale wind turbines and hybrid solar and wind systems, Missouri Wind and Solar offers of wealth of helpful articles, podcasts, and videos to get you started.

My Pet Chicken

www.mypetchicken.com

Should you want to purchase a kit or ready-made coop instead of building one from scratch, check the options on this site. It also offers sources for live chicks and pullets and a great selection of chicken products.

NCAT Sustainable Agriculture Project

www.sustainableagriculture.net

This site from the National Sustainable Agriculture Coalition focuses on sustainable and organic production methods for traditional produce and introduces alternative crops and enterprises.

National Gardening Center

www.garden.org

In addition to a world of attractively presented information, this site provides helpful regional updates as the seasons change, as well as valuable background on edible landscapes.

Omlet

www.omlet.us

Along with its stylish range of manufactured chicken houses, rabbit houses, and beehives, this site provides helpful information on breeds and raising techniques.

Outdoor Water Solutions

www.outdoorwatersolutions.com

Harness wind power to aerate your pond or pump your well with one of the mini-farm-scale windmills from Outdoor Water Solutions. The site also offers a range of electric and pneumatic pumping and aeration products.

Index

A

accessible gardens and paths, 44–45
accessible raised beds, 44
aeration windmill, installing a pump or, 212–17
aquaponics (AP), 226–31. *See also* hydroponics
    about: getting started tips, 229; overview of, 226
    advantages of, 226, 228
    aptitude needed for, 230
    building a system, 228–29
    compared to hydroponics, 226
    cost considerations, 229, 230
    fish types, growth rates, and usefulness as food, 231
    fish waste and, 228, 230
    gravel grow medium, 231
    issues to consider, 230
    managing, time considerations, 230
    plumbing set-up, 228–29
    sound considerations, 230
    water requirements and management, 231
    weight of water and, 228
arbor, 49–59
arbor, trellised, 60–66

B

backyard-homestead shop, 210–11
beehives, building
    Langstroth beehive, 236–43
    top-bar beehive, 245
    Warré beehive, 244
bottom-watered container garden, making, 36–39
building codes, ordinances and permits, 78, 158

C

chickens
    about: cleaning coops, 125; egg production and lighting, 267; needs and housing guidelines, 122
    A-frame tractor for, 142–51
    cooling coops, 156–57
    coop and run for, 124–41
    cutting screening/roofing for, 139–40

electrical outlets/fixtures in coops, 155, 267–68
    extreme cold solutions, 152–55
    extreme heat solutions, 156–57
    finishing touches for coops, 141
    food and water dispensers, 155
    frame materials for projects, 122
    free coop plans, 133
    gathering eggs, 125
    hauling, 151
    herding hens, 149
    insulating coops, 153–55
    misters for coops, 157
    nesting boxes for, 133
    painting coops or not, 139, 151
    PVC hen pen, 111–18
    PVC hurdle, 119–21
    supplemental lighting for, 267–68
    ventilation for, 125, 156
children, involving, 197
codes/ordinances and permits, 78, 158
come-alongs, about, 96, 98, 211
concrete
    about: block size/style options, 20; setting blocks, 22
    raised bed, 19–26
conduit and cable
    about: overview and general guidelines, 260
    cable type, 260
    conduit sweep attachment, 265
    conduit types and fittings, 263
    outdoor electrical boxes and, 261
    running/installing, 260–61, 263–66
    stand-alone receptacle installation, 266
    trenching for, 264, 265
    types of conduit, 260
    UF cable, 260, 263, 264–65
    underground depth requirements, 261
    weatherproof box and receptacle/switch installation, 262–63
    weatherproof boxes, 261, 262–63
container gardens, bottom-watered, 36–39
corner joints, options, 95
cucumber trellis, 67

D

deer, discouraging, 91
doors
    building and hanging, 192–93
    installing, 186
    rough openings, 164
    trimming, 187–89
drywall T-square, about, 211

E

easements, 78, 158
electric fence, solar powered, 106–10
electrical wiring. *See* conduit and cable; wiring, outdoor

F

fasteners
    exterior, 18
    pneumatic option, 130
    Simpson Strong Tie connectors, 49
fence feeding, 204
fences. *See also* posts and postholes
    about: building a good fence, 78; codes and ordinances, 78; corner joint types/options, 95; decorative gate ideas, 103; overview of projects, 78–79
    adding gates, 100–105
    chicken containment (*See* chickens)
    deer defense, 91
    gate latches, 93
    horizontal-board, 88–89
    picket, 86–87
    picket gate, 94–95
    rail-connection, 85
    solar-powered electric, 106–10
    stretching, 96–99
    vertical-board, 90–91
    wood-and-wire, 91–92
fish waste, aquaponics and, 228, 230. *See also* aquaponics (AP)
framing walls. *See* saltbox shed; walls
freeze-proof plumbing. *See* plumbing

G

garden structures
    about: exterior fasteners for, 18; overview of projects, 10–11; pressure-treated lumber and, 13; structures for climbers, 10

accessible gardens and paths, 44–45
arbor, 49–59
arbor, trellised, 60–66
bottom-watered container garden, 36–39
concrete block raised bed, 19–26
cucumber trellis, 67
grow-light stand, 71–74
inclined planter, 46–48
irrigating rooftop raised bed, 27–29
keyhole garden, 30–35
soil blocks, 75–77
tool-storage rack, 68–70
vertical planter, 40–43
wooden raised bed, 12–18
gates
about: decorative touches, 103; handmade hardware alternative, 105; hinges, 104, 105
adding, 100–105
latch selection, 93
picket (including frames), 94–95
sliding latch, 93
goat shed, 194–205
about: overview of building, 194; ready-made kits, 203
fence feeding and, 204
floor framing, 196
goat shelf, 203–4
manger for, 205
materials list, 194
plan, illustrated, 195
roof supports and construction, 200–202
roofing install, 202
wall construction, 197–200
windows, 204
grow-light stand, 71–74

H
horizontal-board fence, 88–89. *See also* posts and postholes
hose shears, about, 211
hurdle, PVC, 111–18
hydrant, freeze-proof, yard installation, 255–56
hydroponics, 226, 232–35. *See also* aquaponics (AP)
about: overview of, 226; understanding, 232

advantages of, 226, 232
compared to aquaponics, 226
cost of system, 232
investment requirements, 232
planting seeds in, 235
propagating plants in, 235
space efficiency with, 232
system assembly, 233–35
time required for, 232

I
inclined planter, building, 46–48. *See also* vertical planter
International Building Code (IBC), 158

J
joints
PVC, 112
wood, corner joint types/options, 95

K
keyhole garden
about: origins of, 30
best vegetables for, 30
building, 30–35

L
Langstroth, Lorenzo, 236
Langstroth beehive, 236–43
light stand, for grow-light, 71–74
locking pliers, about, 211

lot coverage, 158
lumber. *See also* specific projects
anti-rot protection, 131
framing members (illustrated), 161
off-gassing and sealing, 151
painting or not, 139, 151
pressure-treated, 13
priming, 188
recycling, 68
wood shingles, 207

M
manger, 205
mesh
PVC hen pen with, 111–18
PVC hurdle with, 119–21
welded- or woven-wire fences, 91
metal roofing, 208–9

N
nesting boxes, 133. *See also* chickens
notching posts, 84–85

O
ordinances/codes and permits, 78, 158

P
painting wood or not, 139, 151
pathways, accessible, 44–45
pavers, 45, 48
permits and codes/ordinances, 78, 158

Index

picket fence, 86–87. *See also* posts and postholes

picket gate, 94–95

planter, inclined, 46–48

planter, vertical, 40–43

plumb, checking for, 163

plumbing

  about: freeze-proof outdoor spigots, 251; overview of outdoor spigots (aka hose bibbs, sillcocks, and bibcocks), 248; protecting from freezing, 251; water where you need it, 251

  anti-freeze spigot installation, 252–54

  aquaponic system, 228–29

  avoiding water contamination, 251

  backflow/back siphonage and, 251

  drainage solution, 256

  finding frost level and, 255

  freeze prevention, 252, 253

  freeze repair, 252, 254

freeze-proof yard hydrant installation, 255–56

  insulating lines, 253

  sweating pipes, 25

  tunneling under obstacles for pipe, 256

pocket knife, about, 211

posts and postholes

  bevel-cut posts, 85

  foundations and installing posts, 80–83

  installing posts, 81–83

  installing/setting, 63–64

  marking finished height, 63

  notching posts, 84–85

  posthole layout, 61–62

  post-top variations, 87

pressure-treated lumber, 13

projects. *See also* garden structures; *specific items*

  about: author's early farm project observations and, 6–7; overview of, 8

getting started tips, 9

PVC hen pen, 111–18

PVC hurdle, 119–21

R

rafters

  cuts for, 167

  estimating size needed, 167

  gable layout, 167

  for trellised arbor, 64–65

raised beds

  about: exterior fasteners for, 18; pressure-treated lumber and, 13; the quickest raised bed, 17; setting concrete blocks, 22; soil mix for, 26

  best accessible, 44, 45

  concrete block, 19–26

  irrigating rooftop raised bed, 27–29

  wooden raised bed, 12–18

resources, additional, 269

rise and run, roof, 165, 166, 167

roofs, constructing, 165–67

  alternatives, 206–9

  framing saltbox shed (*See also* saltbox shed)

  gable roofs, 165

  harmony of roof and, 166

  metal roofing, 208–9

  rafter size estimation, 167

  rise, run and, 165, 166, 167

  roll roofing, 206

  slope and, 165, 166, 167

  span and, 165, 166

  terms defined, 166

  trusses, 165, 166

  wood shingles, 207

rooftop beds

  irrigating raised bed, 27–29

  soil mix for, 29

S

saltbox shed, 168–93

  about: design basics, 169; overview of building, 168–69; perspective view (components illustrated), 168

  foundation, 170–72

  framing roof, 169, 176–79

framing walls, 172–75, 180–81
materials list, 169
sheathing walls and roof, 182–83
shingling roof, 184–85
siding, 190–91
trimming (soffits, fascia, windows/doors, corners), 187–89
window installation, 186
setback, 78, 158
sheds. *See also* goat shed; saltbox shed; walls
about: overview of projects, 158–59
backyard-homestead shop, 210–11
building basics, 160–67
codes and permits, 158
door construction, layout, and hanging, 192–93
getting kids involved, 197
roof construction, 165–67 (*See also* roofs, constructing)
setback, easements, and lot coverage, 158
windows and doors, 164, 186
shop, backyard-homestead, 210–11
siding, installing, 190–91
sledgehammer, baby, 211
sliding gate latch, 93
slope, roof, 165, 166, 167
soil
blocks, making, 75–77
growing plants without (See aquaponics (AP); hydroponics)
rooftop bed mix, 29
solar power
about: how it works, 218
hybrid (wind/solar) systems, 225
installing, 218–22
solar-powered electric fence, 106–10
span, roof, 165, 166
stretching fences, 96–99
studs, types of, illustrated, 161

T

tools
backyard-homestead shop for, 210–11
storage ideas, 211
surprising, 211
tool-storage rack, 68–70

top-bar beehive, 245
trellises
cucumber trellis, 67
trellised arbor, 60–66
trenching, for conduit, 264
trusses, 165, 166

V

vegetables
best for keyhole garden, 30
bottom-watered container garden and, 36
planting seeds in hydroponic system, 235
soil blocks for starting plants, 75–77
vertical planter, making, 40–43. *See also* inclined planter
vertical-board fence, 90–91. *See also* posts and postholes

W

walls. *See also* saltbox shed
about: general guidelines for building, 161
assembling, 161
bracing, 163
checking for plumb, 163
framing members (illustrated), 161
laying out, 162
raising, 162–63
soleplate layout, 162
squaring, 161
stud configurations at corners, 163
top plates, 163
Warré, Abbé Emile, 244
Warré beehive, 244
water
bottom-watered container garden, 36–39
growing plants in (See aquaponics (AP); hydroponics)
irrigating rooftop raised bed, 27–29
misters for chicken coops, 157
well, drilling your own, 217
windmill for pumping, 214 (*See also* wind power)
well, drilling your own, 217
wind power
about: how systems work, 224–25

anticipating power needs, 223
cylinder pump and, 214
hybrid (wind/solar) systems, 225
pump or aeration windmill, 212–17
for pumping water, 214
storing power, 223
turbine systems, 224–25
turbines for electricity, 222–23
wind availability and, 222–23
windmill perspective view, 214
windows
goat shed, 204
installing, 186, 204
rough openings, 164
trimming, 187–89
wire fences. *See also* posts and postholes
solar-powered electric fence, 106–10
stretching, 96–99
wire-and-wood, 91–92
wiring, outdoor. *See also* conduit and cable
about: extension cords and, 248; overview of, 248
adding lighting with timer for poultry, 267–68
electrical box/receptacle requirements, 262
GFCI receptacle installation, 259
GFCI receptacles explained, 257–58
outlets in coops, 155
power and light for backyard shop, 210
receptacle installation, 262–63, 266
solar power installation, 218–22
switch installation, 262–63
wood-and-wire fence, 91–92. *See also* posts and postholes

Credits

Credits

Page 10: Dan Stultz, Stultz Photography, also **page 11:** *upper left;* **pages 12–29:** Ben Toht, © Greenleaf Publishing, Inc.; **pages 30–35: xx; pages 36–39:** Adam Matthews; **page 44:** © Micka|Dreamstime; **page 45:** *top left,* © Creativethetide|Dreamstime, *bottom left,* © Sutichak|Dreamstime, *top middle,* © Sergei Shchepankevich|Dreamstime, *bottom middle,* © Brandon Bourdages|Dreamstime, *top right,* © Scottnodine|Dreamstime, *bottom right,* © Thanate Rooprasert|Dreamstime;; **49–55:** Donna Chiarelli|CH; **Page 83:** *upper right,* Dan Stultz, Stultz Photography; **page 81:** Brian C. Nieves|CH; **pages 82–84:** all Freeze Frame Studios|CH; **pages 85–90:** Brian C. Nieves|CH; **page 91:** Freeze Frame Studios|CH; **pages 92–96:** Brian C. Nieves|CH; **pages 107–111:** Dan Stultz, Stultz Photography; **page 121:** *upper right,* RGBStock; **page 136:** *upper left inset, upper right,* Rebecca Anderson © Greenleaf Publishing, Inc.; **page 137:** *upper left, lower left,* Rebecca Anderson © Greenleaf Publishing, Inc.; **page 138:** all but *upper right, lower left,* Rebecca Anderson © Greenleaf Publishing, Inc.; **page 139:** all but *lower right,* Rebecca Anderson © Greenleaf Publishing, Inc.; **page 140:** *lower image in sidebar,* Rebecca Anderson © Greenleaf Publishing, Inc.; **pages 155–157:** Brian C. Nieves|CH; **page 159:** John Parsekian|CH; **page 161:** courtesy of Better Barns; **pages 163–188:** Donna Chiarelli|CH; **pages 201–203:** John Parsekian|CH; **page 204:** Dan Stultz, Stultz Photography; **pages 207–212:** courtesy of Outdoor Water Solutions; **page 212:** *box,* courtesy of Mike Willis; **pages 213–216:** Dan Stultz, Stultz Photography; **page 217:** *top,* courtesy of U-Gro Hydroponic Garden Systems, LLC; *bottom,* courtesy of The Aquaponic Source Inc.; **page 218:** courtesy of The Aquaponic Source Inc.; **page 220:** *upper right,* courtesy of The Aquaponic Source Inc.; **page 221:** courtesy of The Aquaponic Source Inc.; **page 222:** © Micka|Dreamstime; **page 223:** © Laura Stone|Dreamstime; **page 225:** © Anastasiia Malinich|Dreamstime; **page 226:** *top,* Dreamstime; *second from top,* courtesy of Growstone; *third from top,* © Opasstudio|Dreamstime.com; *fourth from top,* © Renegadewanderer|Dreamstime.com; *fifth from top,* © Robhainer|Dreamstime.com; *sixth from top,* © Amwu|Dreamstime.com; **pages 226–226:** courtesy of U-Gro Hydroponic Garden Systems, LLC; **page 230:** *upper right,* Dave Toht, © Greenleaf Publishing, Inc.; **pages 238, 241:** Merle Henkenius|CH; **page 242:** John Parsekian|CH; **pages 243–244:** Merle Henkenius|CH; **page 245:** *top,* Merle Henkenius|CH; *bottom,* Dave Toht, © Greenleaf Publishing, Inc.; **pages 246–255:** Brian C. Nieves|CH; **page 256:** *bottom left and right,* Rebecca Anderson © Greenleaf Publishing; **page 257:** Rebecca Anderson © Greenleaf Publishing. All others by Dave Toht, © Greenleaf Publishing, Inc.